P. Albert Morrow

Syphilis and Marriage

Lectures Delivered at the St. Louis Hospital, Paris

P. Albert Morrow

Syphilis and Marriage
Lectures Delivered at the St. Louis Hospital, Paris

ISBN/EAN: 9783337163587

Printed in Europe, USA, Canada, Australia, Japan

Cover: Foto ©berggeist007 / pixelio.de

More available books at **www.hansebooks.com**

SYPHILIS AND MARRIAGE.

LECTURES DELIVERED
AT THE ST. LOUIS HOSPITAL, PARIS.

BY

ALFRED FOURNIER,

PROFESSEUR A LA FACULTÉ DE MÉDECINE DE PARIS, MÉDECIN DE L'HÔPITAL ST.-LOUIS,
MEMBRE DE L'ACADÉMIE DE MÉDECINE.

TRANSLATED BY

P. ALBERT MORROW, M.D.,

PHYSICIAN TO THE SKIN AND VENEREAL DEPARTMENT, NEW YORK DISPENSARY;
MEMBER OF THE NEW YORK DERMATOLOGICAL SOCIETY; MEMBER OF
THE NEW YORK ACADEMY OF MEDICINE.

NEW YORK:
D. APPLETON AND COMPANY,
1, 3, AND 5 BOND STREET.
1882.

TRANSLATOR'S PREFACE.

THE mere announcement of the title of this book is a sufficient apology, if any were needed, for bringing it before the medical profession of this country.

There is not, that I am aware, any work in the English language which treats of the relations of syphilis with marriage. This important subject has been entirely ignored, or only incidentally alluded to, in a majority of our standard text-books on Venereal.

Reference has been made to it in a few special works and monographs not accessible to the general practitioner, but nowhere has it been exhaustively studied, and the practical questions, medical and social, growing out of it, been discussed in a thoroughly systematic and comprehensive manner.

And yet, it will be generally conceded, there is scarcely a subject in the entire domain of medicine of greater practical importance to the profession and to the public, not only on account of the nature of the pathological questions presented, but also on account of the family and society interests involved, and which it is the physician's manifest duty to protect. In considering the social aspects of the question, the author gives special prominence to the moral obligation imposed upon the physician as regards public prophylaxis, and formulates a complete system of rules to guide his conduct in dealing with the various complex and difficult social problems which may arise.

As intimated above, I was led to prepare a translation of this work because it supplies a want long recognized in medical literature, and because the reputation of M. Fournier as a universally acknowledged authority upon the subjects of which he treats is a sufficient guarantee of its excellence and scientific value. To specialists in this department, and others familiar with the advances made in syphilography during the last twenty years, Professor Fournier needs no introduction. His long connection with the Lourcine, the St. Louis, and other special hospitals of Paris, has placed at his disposition rich stores of clinical experience, which his rare discrimination and practiced powers of analysis have enabled him to utilize to the best advantage. His numerous contributions to the study of syphilis, notably among which may be mentioned his studies of syphilis of the brain and the spinal cord, are all stamped with originality and great clinical judgment, and rank among the most valuable acquisitions of contemporaneous medical science. Quite recently, a special chair has been created for him in the Faculty of Medicine of Paris—an unusual distinction, and one which shows the high estimation in which his talents and work are regarded.

The present work exhibits a profound knowledge of the subject in all its relations, united with a rare skill and tact in treating the delicate social questions necessarily involved in such a line of investigation.

The very full synoptical table of contents which prefaces this work renders an index unnecessary, and I have thought best to dispense with it.

NEW YORK, *October 1, 1880.*

CONTENTS.

CHAPTER IV.

PATERNAL HEREDITY.

CHAPTER V.

MIXED HEREDITY.

CHAPTER VI.

MATERNAL HEREDITY.

CHAPTER VII.

PERSONAL DANGERS OF THE HUSBAND.

CHAPTER VIII.

CONDITIONS OF ADMISSIBILITY TO MARRIAGE—ABSENCE OF ACTUAL SPECIFIC ACCIDENTS—ADVANCED AGE OF THE DIATHESIS.

CHAPTER IX.

PROLONGED PERIOD OF IMMUNITY—NON-MENACING CHARACTER OF THE DIATHESIS.

CHAPTER X.

SUFFICIENT SPECIFIC TREATMENT.

CHAPTER XI.

THE USE OF SULPHUR-WATERS—CONCLUSIONS.

PART II.

AFTER MARRIAGE.

CHAPTER XII.

HUSBAND SYPHILITIC AND WIFE HEALTHY.

CHAPTER XIII.

HUSBAND SYPHILITIC, WIFE HEALTHY BUT ENCEINTE.

CHAPTER XIV.

HUSBAND SYPHILITIC AND WIFE RECENTLY CONTAMINATED.

CHAPTER XV.

HUSBAND SYPHILITIC—WIFE SYPHILITIC AND ENCEINTE.

CHAPTER XVI.

DANGERS TO SOCIETY—SOCIAL PROPHYLAXIS.

NOTES AND ILLUSTRATIVE CASES.

SYPHILIS AND MARRIAGE.

INTRODUCTION.

GENTLEMEN : I propose to broach before you, in a series of lectures, a question which, both in a medical and social point of view, is of the most grave and important character, viz., the study of syphilis in its relations with marriage.

This question is eminently complex, as you may judge from its mere announcement. It embraces a multitude of diverse problems, problems difficult, delicate, and perilous, which affect the dearest interests of families, and involve the heaviest responsibility for the physician.

It is my desire and ambition, if not to solve all these problems, at least to state and discuss them before you in a manner to convince you both of the extent of your duty to society in this respect and of the important protective service you have in your power to render it.

A very natural division of the subject presents itself here, viz. : 1st. A syphilitic subject wishes to marry, and comes to consult us in relation thereto. What conditions ought he to fulfill, medically, in order that we may be justified in permitting him to marry? Or, conversely, in what conditions will it be our duty to defer or even to absolutely interdict the marriage?

2d. The marriage is consummated, and syphilis introduced into the conjugal bed. What medical indications are then to be fulfilled in order to lessen or avert the dangers of such a situation?

In other words, what is, what should be, in this case the rôle of the physician either *before* or *after* marriage? Such is the twofold question which we have now to consider.

PART I.

BEFORE MARRIAGE.

CHAPTER I.

PRELIMINARY QUESTIONS.

It will happen to you often, gentlemen, in the course of your practice, to see a patient, known or unknown to you, present himself in your office, who, with countenance preoccupied, almost anxious, will address you as follows: "Doctor, I am contemplating marriage. I have not always been discreet in my bachelor life, and, what is worse, I have not always been fortunate. I contracted syphilis at such a time. I have had such and such accidents. I have been treated in such and such a manner. The matter is now a serious one for me. I have come to ask you if I am thoroughly cured, and if I can, without danger for my wife, without danger for my prospective children, contract the union which I propose. Be good enough to examine me, to interrogate me, and to give me your opinion upon this subject." Now, when such a request is addressed to you, gentlemen, do not misapprehend the gravity of the question. Your response involves interests the most serious, the most sacred, the most dear to the heart of every honorable man, of every respectable family, as well as interests the most diverse and the most multiplied. By this opinion which you are about to formulate, you incur a responsibility which I can not otherwise characterize than as *considerable;* and I do not think I exaggerate in saying that in the province of the physician there are few problems to solve, on the one hand so grave, and on the other hand so complex, so difficult, so delicate, as this.

Judge for yourselves, now, and see what might be the consequences of an *error* committed by you in such a situation. Suppose that the physician pronounces inconsiderately under such circumstances, and falls into one or other of the only two errors possible to commit in the case. What deplorable results does he not bring upon his.patient !

First Hypothesis.—Take the case of a patient who, although having had syphilis formerly, is, nevertheless, by reason of the treatment he has undergone and the present state of the diathesis, in a condition to contract marriage. The physician consulted in this case *mistakes* the situation of his patient, and forbids his marriage. Consequence : Here is a man wrongly condemned to celibacy, banished from the virtuous life which he proposed to enter, and relegated to an irregular life, with all the miseries, social or otherwise, which it entails. Here is a man whose future and whose heart are both broken by a medical decree that forces him to renounce a union likely to assure his position and his happiness ; here, at all events, is a man deprived of family life, deprived of those two things which, after the turbulence of the first years of foolish youth, become the objects of natural and universal aspiration, viz., home and children.

Second Hypothesis.—The error is committed in an opposite direction, and the physician prematurely sanctions the marriage of a man whose syphilis is still active and dangerous.

Then, indeed, the consequences of such a mistake are truly disastrous and ruinous, for :

1st. This man may infect his wife. And what is more deplorable than to give a virtuous young woman the pox as a wedding present !

2d. This infected couple will engender children that will, inevitably, either die almost as soon as they are conceived, or be born with the father's disease. And what more hideous for a young household than the pox in the cradle !

To say nothing of other possible consequences of such a situation, such, for example, as the infection of the nurse, etc.*

Take my word for it, gentlemen, I have witnessed many scenes, many *dramas* of this kind, and, I declare, I know of no position more heart-rending, more lamentable, more atrocious than that of a man who has introduced the pox into his little household—than the situation of this man (1st) in regard to his disconsolate, weeping wife, whose tears are not even accompanied with recriminations or complaints, for love and affection readily forgive ; (2d) in regard to a new family that will not pardon, that has the right to be severe, and exercises that right ; (3d) in regard to the infant which miserably vegetates, and, in place of being the beautiful child dreamed of by the relatives and mother, is to every one, even to the nearest kin, but an object of disgust and horror ; (4th) finally, in regard to an infected nurse, who threatens, who makes scandal,

---

* In another series of lectures, I have studied the serious consequences of infantile syphilis in relation to nurses.—*Nourrices et nourrisons syphilitiques.* Paris, 1878.

2

who divulges, who throws disgrace upon the family, etc. Picture to yourselves such a scene, gentlemen, and judge of the regret, of the martyrdom, of the man who has caused such afflictions.

Now, it is situations of this kind which you, as physicians, are privileged to avert, thanks to your art, your experience, your authority. It is situations of this kind which it is necessary for you to take in at the first glance, when a syphilitic patient comes to consult you upon the admissibility of marriage ; and it is these which you should have present in your mind when, in one way or the other, you formulate your response—let us say, rather, when you pronounce your sentence. Judge, then, from these few words, gentlemen, on the one hand, what an important and exalted post is yours in your capacity as arbiters in such a matter ; and, on the other hand also—I again insist upon this point—what responsibility you incur for the results that may follow. Judge what service you are called upon to render your patient in either alternative : whether you permit him to marry, if expedient ; whether, on the contrary, and better still, you enlighten him upon the possible consequences of his condition (consequences which perhaps he is ignorant of, or which, at least, he but imperfectly comprehends), and thus preserve him from the frightful predicament in which he was about to involve himself.

And, again, note this : At the moment when you render your decision, it is not alone the patient's interests that you hold in hand ; your protective office extends beyond him, and reaches farther. For behind that patient there is a young wife, there are children yet unborn, there is a family, there is society itself, to be shielded at the same time by your prohibition. Behold, then, how extended and elevated the medical adviser's mission becomes when he finds himself thus the arbiter of so many united interests.

Before entering into the heart of this question and discussing the different problems which will be the object of this exposition, let us first state a principle by endeavoring to define clearly the situation, as it presents itself in practice, and determining exactly the rôle which is assigned us.

When a patient comes to request our advice, to ask our counsel upon the propriety of marriage, despite syphilitic antecedents, it is *as physicians*, as physicians exclusively, that we are consulted. Our rôle is distinctly determined by that fact alone. It is *as physicians* that we are to respond. In other words, it is a question of pathology, and of that only, that we are called upon to judge, and our duty, our absolute duty, is to decide it upon pathological grounds alone, without allowing ourselves to be influenced by any other considerations, whatever they may be.

But, you ask me, wherefore this preface? Why this rule of conduct laid down at the opening of the exposition?

In practice, gentlemen, it is necessary to look upon things as they exist, and to take men for what they are. Now, learn this, presuming that you may not know it already: among the numerous patients who will come to consult you upon the propriety of marriage in the special conditions which now engage our attention, there are many, assuredly (let us even say a very large majority), who will present themselves to you with the twofold intention of learning precisely how to act in their situation, and of submitting themselves to the decision which you will pronounce ; that is to say, of renouncing a projected marriage, if you prohibit them from marrying. This class you do not have to distrust. But there are others of an altogether different kind ; and these last, more numerous than you would suppose *a priori*, will come to you with a tacitly formed resolve to act solely according to their inclination, whatever you may say to them, and to marry at all hazards, and in spite of all your interdictions, because it pleases them to marry, because they have resolved to marry, before even crossing your threshold.* In these

---

* I have seen numbers of syphilitic subjects marry in opposition to all medical prohibitions.

One can not even form an idea of the disdainful indifference which certain people profess with regard to medical prescriptions concerning the subject under consideration. . . .

One of our most learned and esteemed *confrères*, Dr. X., was consulted by a young man, the son of a family of his most intimate friends, for accidents of secondary syphilis consecutive to a recent contagion. Knowing that the young man was then contemplating marriage, he took occasion to accompany his prescription with a long moral upon the dangers of syphilis in relation to marriage, and endeavored to obtain from his patient a formal renunciation of all matrimonial projects. . . . In response, he received, some weeks later, a letter announcing the marriage, with an invitation to be present at the ceremony, celebration, dinner, etc.

It is useless to speak of the reception given to this ironical missive by our *confrère.*

But the expiation could not be long delayed, and it was severe, as you will see.

Three months later, the young couple presented themselves to Dr. X., under the pretext of a "*visite de noces.*" After the usual compliments, the young husband suddenly changed the conversation, and requested some medical advice for his wife, who, in the first place, presented the earlier symptoms of pregnancy, and who, in addition, had had for some weeks a "slight hardness " on the lip. Now, upon examination, this was found to be nothing else than a syphilitic chancre—a chancre manifestly transmitted to the young wife from her husband, who had not ceased for some months to be affected with buccal syphilides, and was still affected with them at that time.

It is almost superfluous to add that the chancre was followed by the accidents of constitutional syphilis. In addition to which, eight months later, the young wife was delivered of a poor, weakly, insignificant infant, which was soon covered with syphilides, and quickly succumbed.

Another example of the same kind:

A young man contracts syphilis and comes to seek treatment from me. Some months later, still affected with secondary accidents, he announces to me that he

cases, it is not advice, not counsel which they expect of you; it is your consent, your acquiescence in their projects which they hope to wring from you. Such consent, indeed, would set them at ease with their consciences, and, moreover, should things happen to "turn out badly" in the future, it would serve them as an excuse and an exoneration.

Now, in order to gain their ends, in order to force your convictions, those of this latter category, while pretending to consult you, never fail to shirk the medical aspects of the question almost at once, in order to carry you off

finds himself engaged, "almost in spite of himself," to be married, which must necessarily take place quite soon. I energetically insist upon his renouncing such a project; I depict to him the dangers to which he is about to expose himself and his future family. I endeavor to convince him of the immorality, of the culpability of such an act, etc. Nevertheless, he marries, and I see him no more for a certain time.

Some months later he comes to me in a veritable state of affright and distress. He has infected his wife, he tells me, and he comes to ask my attentions to her. I find, in reality, this young wife in an active condition of syphilis. I prescribe a treatment, a hygiene, etc., and, especially, I expressly recommend the husband to avoid at any price the possibility of pregnancy in such a situation. I explain to him, superabundantly, that a pregnancy would be a second disaster, for it could only end, according to every probability, either in an abortion or in the birth of a syphilitic infant.

Notwithstanding, two months later, the young wife becomes *enceinte*. I then treat her so much the more energetically, and I have the happiness to prevent an abortion. Then, when I consider myself sure of obtaining an accouchement at full term, I announce the express absolute obligation for the mother to nurse her child. At least, I say to the husband, arrange it so as to avoid a third misfortune. Do not confide your infant to a nurse, for it is quite probable that the nurse will receive the pox from it. M. Ricord, consulted upon this point, confirms the apprehensions expressed by myself, and energetically insists upon the absolute necessity of the mother nursing the child.

Some months pass away, without my again seeing this family. Then, one day, the father reappears, bringing the child covered with syphilides, and the nurse to whom this child had been confided. As I had foreseen, this nurse had become infected, and bore upon one of her breasts an indurated chancre of the most typical character.

To recapitulate: triple transgression of medical advice and triple disaster, viz., the infection of a young wife, the birth of a syphilitic infant, and the contamination of a nurse.

to considerations of a totally different character. They have a hundred pretexts at their service to plead their cause and to induce you to share their feeling. One, for example, "has given his word, he is engaged, formally engaged, and you should not compel him to break his plighted faith."

Another will invoke an urgent material necessity. Counting upon the dowry of his future wife, he has just purchased an office, a practice, a commercial business, or the like. "If you force him to break off the marriage, it will be to bring him to ruin, failure, and dishonor."

Still another, more crafty, will work upon your feelings. "I love a young lady," he will tell you, "and am beloved of her. Our two families, our aged parents, place their dearest hopes on this marriage, and a rupture would break all our hearts," etc.

All these arguments (copied from life, and reproduced literally, on my honor)—all these arguments, and many others which I pass in silence, have nothing to do, gentlemen, with the situation which we are called upon to examine as physicians. Were such reasons as valid as they are detestable, we have nothing whatever to do with them, nor have they any value for us medically. Let them be null and void for us. Let us stand firm under such circumstances and reject everything that does not bear upon the clinical aspect of the case. Let us not quit our own ground, but confine ourselves to a strictly pathological view of a question which for us should not be separated from pure pathology.*

I will even go farther, and assert that we should be culpable, veritably culpable, were we to do otherwise;

* Cf. Langlebert, *La syphilis dans ses rapports avec le mariage.* Paris, 1873, p. 10.

that is to say, were we to allow ourselves to arrive at a decision based upon considerations foreign to our art. And the proof of this is the embarrassment which we should have in justifying our. conduct if a misfortune should occur, if a patient, whom we had permitted to marry from extra-medical considerations, should happen to introduce syphilis into his family. What plea could we then urge against such an accusation thrown in our face as, "What! you judged this man to be medically unfit for marriage, dangerous for marriage, and, because he has alleged questions of convenience, of position, of pecuniary wants, of feelings, you authorized him to run the risk of introducing the pox into his home!"

Let us guard against the possibility of such recriminations, let us guard against a grave fault into which (I can say from experience) we are but too easily led by an excess of kindness ; the more so, as we have a ready means of avoiding it, viz., not to deviate from the rôle naturally assigned to us by our profession.

In a word, consulted under such circumstances, let us keep to medicine, and judge only as physicians the case submitted to us. Let there be no concessions to arguments of a foreign nature ; no condescensions which we may bitterly regret in the future, and which, without benefiting any, may prove prejudicial to all, seriously compromising both our authority and our dignity.

With these preliminaries, we shall enter upon the principal subject. But, first of all, let us examine a primary question, which, if resolved affirmatively, would exclude all further discussion, by rendering useless that which is now to follow.

Does syphilis constitute an express interdiction, an absolute obstacle, to marriage?

Without doubt, gentlemen, you have often heard the common adage, "*avec la vérole il faut rester garçon.*" Many people repeat and spread this as an axiom indorsed by all the faculty, who, moreover, in order to discourse more at ease of such things, have never taken the care to study them. It is also affirmed in a much more energetic manner by families (and I excuse them), who have been interested in the question and seen syphilis introduced into their household under cover of marriage. These families can not sufficiently reprobate every man who, with the syphilis, would dare ever to aspire to the position of husband. In their opinion, and in the opinion of all those who have been the victims of similar calamities, the pox is radically "incompatible with marriage."

By many physicians, syphilis would be regarded as an express contraindication to marriage. I have had the pleasure of conversing upon this subject, which has for many years engaged my attention, with numbers of my *confrères*, and many of them have said this to me in their own words : "Nobody marries, no one ought ever to marry, when once he has had the misfortune to contract the pox." I could even cite two of our most esteemed *confrères* who have joined practice to precept, who have exemplified it by remaining unmarried, from the sole consideration that they had acquired syphilis in their life as students. One of them, a most distinguished physician, whose heart is on a level with his talent, has never allowed himself to be dissuaded by me (who have the honor to be his friend) from that which he terms his "incapacity for marriage." "You have spoken to no purpose," he has repeated to me a hundred times. "When one has the pox, one should keep it to himself, without running the risk of giving it to others, especially to his wife and children." To which I have rejoined, on my part, "When one has the pox, one should cure it; and when by force of care one has rendered it innocuous for others as well as himself, then, having again entered upon a normal condition, one has the moral right to aspire to marriage."

And, in fact, gentlemen, the truth is not with those who would make of syphilis an insurmountable obstacle, a permanent, eternal, absolute interdiction to marriage.

The truth is, that, save very rare exceptions, syphilis constitutes only a *temporary* interdiction to marriage, and that a syphilitic subject may, after a certain stage of sufficient depuration, return to a state of health which fully restores his fitness for the double rôle of husband and father.

Upon this point, I appeal to common, to daily observation. Do we not encounter almost every moment, both in private and hospital practice, persons who, having had syphilis in their youth, afterward marry, and who, married, on the one hand, *have never communicated anything syphilitic to their wives*, and, on the other hand, *have had healthy children*, well developed and thriving, as active and as well favored as could be desired.

Cases of this kind abound and superabound. I deny that any physician engaged in practice for several years can not bring forward his contingent of personal examples to the support of the proposition, so consoling, which I have just enunciated. For my part alone, I have in hand (to speak only of recorded facts) eighty-seven observations relative to syphilitic subjects, undoubtedly syphilitic, who, having married, have never communicated to their wives the least suspicious phenomenon ; and, what is more, have begotten, these eighty-

seven, a total of one hundred and fifty-six children absolutely healthy.*

These observations, which I have chosen among many
others, are the most convincing to me, and they will become so to you, gentlemen, I trust, when I shall have
assured you that they all relate to patients and to families
that I have scrupulously examined and followed up, that
I have had under observation for many years, and many
of whom are still in close relations with me. In addition,
allow me to cite you two of these cases as examples:

Two of my patients, formerly syphilitic, married, one
without consulting me, the other after having received my
consent. The first has to-day *four* children, and the second *five*. Now, after a dozen years that I have been the
physician of these two families, I can affirm to you with
every assurance that (1) the wives of these two patients have
never presented the least suspicious phenomenon, or the
slightest manifestation which had any analogy with syphilis; (2) that the nine children of these two families are
absolutely sound and healthy. Thanks to the solicitude
of their mothers, I have been able to look after them at
leisure, from their birth until the present time, not only
in their various sicknesses, but in the slightest indisposition, with which they have been affected. Never have I
detected in them anything which recalled the paternal disease in any degree or in any form whatever. What more
do you wish? Here, then, incontestably, are two families,
where the syphilis of the father has played no rôle whatever, where things have progressed as they would have
progressed in the absence of any syphilitic antecedent.

One of these two families (pardon me this incidental
digression) presents the perfected type of domestic happi-

---

* V. Notes and Illustrative Cases, Note 1.

ness, and has more than once suggested to me a reflection relative to the subject we are now considering. I have often said to myself, when seated at this happy hearthstone, and a witness of its intimate joys: "What a mistake should I have committed if, from an exaggerated fear of the old disease, I had prevented this marriage! What a mistake should I have committed, if I had nipped in the bud all the present felicity of these two beings so affectionately united—if I had prevented these beautiful children from coming into the world!"

Yes, then, a hundred times yes, *one may marry after having had the pox*, and the results of a marriage contracted under these conditions may be absolutely safe, medically speaking. This I affirm and proclaim boldly, after having carefully studied this serious question from both a clinical and a social point of view; after having conscientiously investigated numbers of cases, personal or contributed by others. This is for me an established fact, a demonstrated verity.

But, this said, I would hasten to add immediately: If one may marry after having had the pox, one may not, one ought not, to marry in this special situation without being declared free from liability to subsequent manifestations of the disease; without satisfying certain conditions which are indispensably necessary. What those conditions are we shall now endeavor to define.

CHAPTER II.

DANGERS DUE TO SYPHILIS IN MARRIAGE—DIRECT
CONTAGION.

In order to determine on what conditions, medically
and morally, a syphilitic subject may be permitted to
contract an alliance, it is necessary, first of all, to ascertain how and in what respects that man may become
dangerous in marriage.

Such is, naturally, the primordial point to establish,
for such is the basis of all reasoning in solving the problem which is now imposed upon us.

Now, in my opinion, and as I understand the question,
a man with syphilitic antecedents who contracts marriage
may become dangerous in marriage in the three following relations: 1st, as husband; 2d, as father; 3d, as head
of the social community constituted by marriage. In
other words, he may become dangerous:

1. To his wife.
2. To his children.
3. To the common interests of his family.

Let us see what this programme signifies, and examine
from every point of view the three terms of the proposition which I have just formulated.

First Point.—*A man, who with syphilitic antecedents
contracts marriage, may become dangerous to his wife.*

This is evident; it admits of no discussion, in fact. It is manifest that a healthy young woman delivered over to a syphilitic man may suffer the *contre-coup* of this syphilis. This is what common sense says *a priori*, and experience confirms it.

And, in reality, how often have we not seen, who has not seen, cases of this kind? A young woman in a perfect state of health marries a man who has acquired syphilis in his bachelor life. Several months later you find her *syphilitic*, and syphilitic through the sole agency of her husband.

This syphilis in young married women—we may say incidentally, since occasion presents—is *quite frequent*, much more frequent than one would suppose. In proof of this, you will find many cases recorded in various reports. In proof, also, the following personal statistics:

In a total of 572 syphilitic women who have come under my observation in my private practice, I found no fewer than 81 who contracted syphilis *from their husbands soon after marriage.* These figures are sufficiently eloquent by themselves to render all commentary unnecessary. Let this be a warning, then, to families who consider not the health of their future sons-in-law, and who neglect to protect their daughters against careless, indifferent, or unprincipled men, to whom it is a matter of little concern to carry the pox into the conjugal bed.

This fact, then, is patent: frequently young married women receive syphilis from their husbands. This fact we must now explain.

How do wives in such circumstances receive syphilis from their husbands? How does a syphilitic husband become dangerous to his wife? How, in a word, is the syphilitic contagion transmitted from the husband to the wife?

In two ways, one of which is very simple, trite, and commonplace, and the other is special, mysterious, not materially demonstrable, but which is undeniable, as has been demonstrated by observation. To explain this, certain preliminary statements are necessary.

The first mode of contagion, which I have just now characterized as the common, ordinary one, consists simply in this : *Transmission to the wife of a contagious lesion occurring in the husband after marriage.*

A syphilitic husband not yet cured of his syphilis comes to be affected with a suppurative lesion of a specific nature. He has connection with his wife at this time. Naturally, he infects this wife from the lesion which he actually has at the time. There is nothing but what is very simple, nothing but what is absolutely normal in this mode of contagion, which is, as every one knows, the ordinary mode by which syphilis is transmitted, propagated, and maintained.

As examples of this kind : A young man marries after fifteen months of syphilis. He comes to have on his glans two circinate erosions of the kind which we term in technical language papulo-erosive syphilides of annular form. Regarding these lesions as herpes (another affection to which he is subject at times), he continues to have intercourse with his wife, and thus communicates syphilis to her, which makes its début by an indurated chancre on the vulva, soon followed by general symptoms.

Another young man, belonging to the highest social circles,* marries despite my advice, after two years of

---

* I note by design this particularity, as I shall again note it if occasion serve. And, in fact, numbers of persons imagine and repeat that transmissions of syphilis in marriage are scarcely ever met with except "among the lowest classes," and as a result of ignorance, of carelessness, of misery, etc. Now, this is a delusion, a grave error, against which daily experience protests. The cases of this character

syphilis. A great smoker, he is often affected with slight labial erosions, to which he pays no attention. Notwithstanding my advice, he obstinately refuses to consider them dangerous or to have them treated. Consequence: From one of these erosions, which I had regarded as undoubtedly syphilitic, he eventually transmits syphilis to his wife, upon whom I afterward discover an indurated chancre of the lower lip.

One of my professional friends contracts syphilis. He treats himself. Thinking himself cured, three years later he marries. Some months afterward I receive a desponding letter from him. "A lamentable catastrophe," he writes, "has befallen me. Quite recently I have had the misfortune to infect my young wife, nineteen years old; and I have infected her—it is not to be believed—with a miserable little papule on the penis, an erosive papule, it is true, but minute, absolutely minute to the degree that I did not at first perceive it, and afterward took no precaution!"

And, in like manner, gentlemen, I could accumulate here many other cases of the same character, differing perhaps in detail, but all identical in nature. This first mode of contagion, I repeat, is frequent, then, even in marriage. And how could it be otherwise, considering the extreme contagiousness of the suppurative form of secondary accidents—considering the repeated developments, so frequent and so easy, of accidents of this order in syphilitic subjects imperfectly treated—considering the multiplicity of

are encountered almost equally in all classes, from the humblest to the most exalted. I declare, for my part, that I have observed a great number in the higher bourgeoisie, even among the aristocracy—that is to say, in social surroundings where education, intellectual and moral culture, absence of pecuniary cares, personal independence, the free gratification of the desires, etc., etc., ought, it would seem, to exclude such shames.

the relations, familiarities, and contacts of every kind
which, in domestic and family life, expose the wife to infection from her husband?

The last consideration which I have just mentioned is
of the first importance, and I beg you to remark it. Contagion, in fact, is rendered so easy by the close and continual intimacy which results from marriage, that it becomes almost inevitable. According to experience, it. is
rare to see a healthy woman live long in contact with a
syphilitic man (or inversely, but the converse does not interest us just now) without the healthy partner becoming
contaminated by the diseased partner. As that witty observer, M. Dechambre, has said: "The pox, like the daily
bread, is divided between husband and wife."

CHAPTER III.

THE second mode in which syphilitic contagion oper-
ates in marriage is entirely different from the preceding, and absolutely special, as you will see. It consists
in what is called *syphilis by conception*. Little known or
at least little believed in by us, denied even by many of
the profession, this syphilis by conception ought to receive
here some consideration, for it constitutes a part, and a
very essential part, of our subject.*

How does it present itself in practice? How does it
manifest itself to observation? As follows: A pure and
healthy young woman is married to a syphilitic man,
whose syphilis is not yet eradicated. Summoned to attend
her some months later, you find her syphilitic; you find
her, for example, affected with unmistakable secondary
symptoms, such as the cutaneous syphilides, buccal mucous patches, acneiform crusts of the hairy scalp, cervical
adenopathies, pains in the head, wandering neuralgias,
lumbago, intermittent febrile attacks, alopecia, etc. There
is no doubt possible: this woman is thoroughly syphilitic.
This settled, you proceed to investigate the why and

---

* If I did not treat this question in a manner exclusively incidental, I should
cite here the opinions and the well-known works of MM. Depaul, Diday, Hutchinson, de Meric, Melchior Robert, Bazin, etc.

3

wherefore of this syphilis. How did syphilis attack this
young wife ; in what way was it introduced ; what was the
initial accident; where was the seat of the chancre, etc.?
And, now, a double astonishment begins for you. There
is, in the first place, no trace of what is called the primary
affection, no vestige of a chancre, no recollection of a
localized lesion having preceded the present symptoms by
several weeks. "Pass over the chancre," you say to yourself, "for every one knows that in the female the chancre
is often but a slight, fugitive lesion, which may easily pass
unperceived by the patient, and can not, after a very short
time, be detected by the physician. But, at least, I shall
find a bubo, for the bubo is not only the 'faithful companion of chancre,' according to M. Ricord, but also a *post-
humous witness* that long survives it—that testifies to its
previous existence long after its disappearance and cicatrization." You then search for the bubo, and you do not
find it either. There is no trace anywhere of a primitive
adenopathy. In a word, there is nothing but secondary
accidents, as if the syphilis had manifested itself *d'emblée*
in this patient—as if it had never had a primary stage. At
the first blush, this seems very strange, does it not? But
this is not all. Another surprise immediately awaits you.

Syphilis thus verified in the young wife, you take the
husband aside, who confesses his syphilitic antecedents to
you, if you are not already aware of the fact. Then you
naturally ask him what *fresh* accidents he has experienced since his marriage which could infect his wife. Upon
this, protestations, strong protestations, on the part of
your patient! "No; I tell you I have had no new symptoms since my marriage ; nothing, absolutely nothing. I
was aware of my condition. I had been warned by my
physician of the dangers which my wife would incur if

there should happen to me any accident similar to those I had already had. I have been on my guard, I have examined myself, I have watched myself scrupulously, and I can affirm to you in the most positive manner that nothing suspicious has appeared upon me since the day of my marriage. That I will answer for."

Not content with these assertions, you proceed to make a careful examination of your patient; the result of the examination is negative. Not the least evidence of the disease is present either upon the skin or upon the mucous surfaces—not the least trace of a lesion recently vanished. So that, judging from appearances, one is forced to admit this : that the young wife has become syphilitic from the contact of her syphilitic husband, *without the husband having at the time the least external lesion capable of infecting her.**

Without doubt, gentlemen, if cases of this kind were observed only once, perchance in a manner altogether exceptional, one would have the right, strictly, to call them in question, to reject them, and to say : "These cases are null and void, incomplete, defective. These are cases where there has been a mistake, either on the part of the woman, who has not seen or felt the chancre, or on the part of the husband, who deceives himself or who deceives us. Let us pass them, then, without attaching any importance to them." But the truth is, on the contrary, that cases of this kind are common ; they are frequent ; they

---

* I do not bring into discussion here the possibility of a contagion by the *sperm.* From ancient date it has been established by clinical observation that the sperm of a syphilitic subject is not susceptible of transmitting the contagion. Experimentation has recently pronounced itself in the same sense. Healthy subjects have been inoculated with sperm derived from syphilitic subjects, and, as might have been expected, the inoculation has been without result.—(V. Mireur, *Recherches sur la non-inoculabilité syphilitique au sperme,* publiées dans les *Annales de dermat. et syphilig.,* t. viii, p. 423 ; Dr. X., oral communication.

obtrude themselves upon our observation with a significant persistence ; they present themselves always identically the same, always under the same conditions ; finally, they force themselves upon us at times, carrying conviction with them.

In reality, this is not a class of cases that can be ignored, or in which it can be objected that we have to do with a negligent husband, unconscious of danger, careless of his person—a poor observer, liable to allow a specific accident upon himself to pass unperceived. These are cases of an entirely different character, which have been recorded of husbands very attentive to the state of their health, scrupulous, conscientious, apprised of the dangers to which their wives are exposed from their old diathesis, and who never cease to examine themselves with the most minute care. There are cases of this kind, finally, which have been observed by physicians *in propriâ personâ*, in their own families. I know of many of them which, unfortunately, I am not privileged to cite.*

Now, when these husbands, who so intelligently appreciate their condition, when these professional men repeat to our satisfaction : "No, I assure you, I have not had any symptom since my marriage ; I have not had, either on the penis, or in the mouth, or elsewhere, the smallest erosion, the least scratch, the slightest abrasion capable of infecting my wife"—when from such assertions we are furnished with such guarantees, and when these assertions are reproduced identically parallel, not only in a few cases, but in a multitude of cases of the same kind—when a fact so inexplicable, so extraordinary, as this may appear, becomes no longer the exception, but the common, habitual, almost general rule—under such circumstances we are

---

* Cf. J. Hutchinson, "Medical Times and Gazette," December, 1876, p. 643.

forced to yield, to surrender an incredulity, otherwise quite legitimate, and to conclude, finally : "It must be so ; here is a woman who, on the one hand, has the syphilis *without having presented an initial lesion*, and who, on the other hand, has been infected by her husband *without the husband presenting any contagious lesion.*" But, then, what is this mystery ?

What, then, is this mystery ? Well, gentlemen, here is the key : That the woman become syphilitic in this manner without initial lesion, without chancre, and become syphilitic from the contact of a husband exempt since his marriage from every contagious lesion — that woman is *enceinte*, and she has received the syphilis by *conception.*

In cases of this kind, gentlemen, there is always a special element which intervenes to complicate the morbid ensemble ; and this new, this supernumerary element (permit me the word) is *pregnancy.* In such a situation, *pregnancy is never absent.* If it happens to you (and it will happen many times in your practice) to encounter a woman who has acquired syphilis without presenting primary accidents, and, moreover, has acquired syphilis from a syphilitic companion who has for a long time been free from every suspicious symptom, direct your attention always to the question of pregnancy ; interrogate, examine the woman from this point of view, and invariably you will succeed in establishing this : either this woman is actually *enceinte* at the moment of your visit or she has been recently *enceinte*, and has just been confined, or she has had a miscarriage.

Ah, then, if this be the case, if always, and invariably, the facts which we are now studying present themselves thus, with the necessary addition of a special element, pregnancy, this fact becomes a ray of light for us.

Since these cases which deviate from the normal laws of syphilitic contagion are always complicated by a special element, which intervenes in a constant manner, may not this element prove to be the cause of the said anomaly? May not pregnancy play a rôle here in determining this apparent modification of the usual modes of contagion? May not this woman, who appears to have contracted the syphilis from her husband, in reality have received it from her child—from the infant which sojourned in the maternal womb—with the syphilis which it received from its father?

Indeed, yes, gentlemen, and such, there can be no doubt, is the origin of the syphilis in the cases in question. The wife-mother, infected thus, that is, become syphilitic without initial accident, and become syphilitic from the contact of a husband long free from every external manifestation, has received the syphilis, not from her husband, but *from her child.*

We have not here a syphilis transmitted by contagion in the usual, habitual manner : there is here a syphilis conceived *in utero*, introduced by the infant into the womb of its mother, communicated to the mother by her infant—in a word, it is what is termed SYPHILIS BY CONCEPTION.

It would be foreign to my subject to enter here into the clinical history of syphilis by conception, so different from ordinary syphilis both in its origin and in its primordial evolution ; but it is important that I leave you in no doubt as to its authenticity, and with this view I add the following considerations :

1. In the first place, if we reject this pathogeny of the infection transmitted to the mother by the fœtus in the class of cases which we are now considering, viz., the

syphilis of young wives who never present the primary accident, chancre, and who receive, or seem to receive, the contagion from a non-contagious husband—this syphilis, I say, will remain absolutely incomprehensible, absolutely inexplicable.

And I repeat this again : the cases of this kind are too numerous, too well authenticated for us to refuse to receive them, for us to think of interpreting them by assuming material errors of observation. They actually obtrude themselves in practice, and it is necessary to recognize them as facts without discussing the theory.

2. These same cases, which deviate from the general laws of ordinary syphilis, never thus deviate except with the addition of a *special* element, which is none other than pregnancy. Always and invariably they occur in women *enceinte* or recently confined. Is not this significant? Does it not imply that pregnancy plays here a *special* rôle in modifying the usual conditions of syphilitic contamination?

There are, moreover, certain facts more conclusive still, if possible. These facts may be summarized thus : A healthy woman is united to a syphilitic man. So long as she does not become pregnant, she remains uncontaminated ; but let her become pregnant, and, immediately, syphilis breaks out upon her.

Now, why this immunity before pregnancy, and why this infection taking place with pregnancy, if conception be immaterial, if it have no part in the specific contamination.*

3. A third argument arises from the morbid condition of the infant. I will explain what happens to the infant

---

* Cases of this order have been signalized already by numbers of physicians. I confine myself to announcing the fact without citing particular examples.

in this class of cases of which we are now speaking.
Most commonly, in truth, it dies before being born. Was
it or was it not syphilitic? We know nothing positively
upon this point, and we have nothing to say, although the
sole fact of its death constitutes a presumption in favor of
syphilis. But in other cases it is born alive, and then it
always manifests unequivocal symptoms of syphilis ; *it is
always syphilitic.* Now, if the infant in such circumstances is tainted with syphilis, what is there impossible
or extraordinary in its transmitting the disease to the
mother during its intra-uterine life?

If maternal syphilis has the power (which every
one admits) of reflecting itself upon the infant, why
should not the syphilis of the infant reflect itself in like
manner upon the mother? What! Here is an infant
which, procreated syphilitic by the agency of its father,
lives syphilitic during several months in the womb of its
mother, and yet you would think it extraordinary, impossible, that the infection of the infant should be transmitted to the mother! A syphilitic organism included
within a healthy organism, and the one not contaminate
the other! In truth, it is not the infection of the mother
under such conditions which would in my estimation constitute the surprising fact; for me, the surprising fact
would be that the mother remained refractory to such
chances of contagion.

4. After all, syphilis by conception is but the analogue
of the syphilis which, in the course of pregnancy, reflects
itself in an opposite direction, from the mother to the infant. It is necessary, then—common sense indicates it—
that it obey the same laws as the latter, and this is precisely
what takes place, as you will see.

The peculiarity of hereditary syphilis, you know, is to

make an invasion *d'emblée* by general symptoms; that is, to have no primary stage, in a word, to be exempt from those two accidents which constitute the inevitable, necessary début of every syphilis contracted in the usual way of contagion, viz., chancre and primitive adenopathy symptomatic of chancre. Now, the syphilis of conception has precisely the same peculiarity. It also admits neither chancre nor bubo into the number of its constituent symptoms. It also makes invasion *d'emblée* by manifestations of a general order, and this deviation from the great laws which govern syphilis in its habitual forms certainly finds its analogue in the special mode which here presides over infection.

Such are, in a very succinct form to be sure, but sufficient I think for our present subject, the considerations of various kinds which establish the undeniable fact of the possible infection of a woman by way of conception. The fact admitted and accepted, it now remains for us to discuss its interpretation, if this does not exceed the limits within which we must confine ourselves. How does syphilitic impregnation irradiate from the fœtus to the mother in the cases we have just considered? Does the maternal infection result from the contact of a fecundated ovule, and propagate itself, either in the Fallopian tubes or in the uterus, at the time when this ovule is not attached to the mother by any organized graft? Or, indeed, is it effected subsequently by the exchanges of the placental circulation?* or does it take place in some other special and unknown manner? Upon this point we confess our complete ignorance. We know nothing of the process, of

---

* This is what, for example, M. J. Hutchinson declares, who has given to this mode of infection of the mother by the infant the expressive term of *fœtal blood contamination.*—V. Memoir cited, "Medical Times and Gazette," 1876.

the mechanism of the infection, and, relative to this point, we can only emit hypotheses without value. Infection takes place only in these special conditions, and often, very often, the wives are the victims. The knowledge of this alone is sufficient for our present purpose. Let us, then, accept the fact, and leave the interpretation aside.

This settled, let us reunite the elements which precede, and summarize the facts relating to our first proposition, by saying: A man with syphilitic antecedents who contracts marriage may become dangerous to his wife in two ways:

1. *Directly* by transmissible *contagious lesions*, which may happen to him after marriage.

2. *Indirectly*, through his *fecundating power;* that is, by the *procreation of an infant, the infection of which may be reflected upon the mother.*

CHAPTER IV.

PATERNAL HEREDITY.

Second Point. — A man with syphilitic antecedents who contracts marriage may become dangerous to his children.

1. Until within a recent period, the theory of the *paternal* heredity of syphilis was accepted without opposition, except from a few. It was not questioned that a syphilitic father could, must even, beget syphilitic children. It was an opinion generally admitted, and science seemed definitely settled upon this point.

But the aspect of this question has indeed changed within late years : numerous observations, important investigations, have sprung up on all sides, tending to singularly restrict the sphere of paternal influence in the hereditary transmission of syphilis.* But this is not all ; some investigators have gone further in this direction.

* Cullerier, *De l'hérédité de la syphilis.—Mémoires de la société de chirurgie de Paris,* 1851, t. iv, p. 230.

Notta, *Mémoire sur l'hérédité de la syphilis.—Archives générales de médecine,* 1860, t. i.

Charrier, *De l'hérédité syphilitique.—Archives générales de médecine,* 1862, t. ii.

Durac (J. E.), *De l'hérédité de la syphilis.—Thèses de Montpelier,* 1866.

Mireur (H.), *Essai sur l'hérédité de la syphilis.—Thèses de Paris,* 1867.

Owre (Adam), *Sur l'étiologie de la syphilis héréditaire.—*Analyse dans les *Annales de dermatologie et de syphiligraphie,* t. v, p. 388.

Sturgis (F. R.), "Notes upon Certain Points of the Etiology of Hereditary Syphilis."—Analysis in *Annales de dermatologie et de syphiligraphie,* 1877, t. ix, p. 113.

They have gone to the point of denying the paternal influence in the transmission of the disease, and of saying: "The influence of the father is null, absolutely null, in the transmission of syphilis to the fœtus. The child of a syphilitic man is born sound, exempt from syphilis, and perfectly healthy."

You can conceive, gentlemen, the importance of this question in connection with the special subject which now engages our attention. For, when a syphilitic patient comes to consult us in order to know whether he may or may not marry, our responsibility will be by so much lightened, if we have before us the certainty that this man, although syphilitic, can in no way be prejudicial to his children. That is self-evident. Let us examine this question, then, with all the care, all the solicitude which it merits. The new doctrines which have been introduced relative to the non-transmission of syphilis by paternal heredity, or, more generally, the non-influence of paternal syphilis upon the offspring—these new doctrines, I say, while containing an element of truth, contain manifest exaggerations, and more than exaggerations, absolute errors, dangerous from a social point of view, dangerous in every respect, and which, in consequence, should be energetically combated. For my part, indeed, according to my own experience, as well as from numerous observations which have been furnished either from my reading or from communications courteously made to me, I regard it as certain that a syphilitic father can, by virtue of a syphilis still recent and active, be eminently prejudicial to his children. And of this I am in a position to furnish the proofs, as you shall see.

In the first place, to consider the matter from a purely theoretical point of view, how can we admit, for a single instant, that the condition of a syphilitic father can be inoffensive to his offspring? What! when we see constantly, and in a manner so evident, paternal heredity manifest itself in the child by so many resemblances of every kind; when we see it attested, not only by physical or moral analogies, but also by the most striking pathological analogies, we are to believe that this paternal heredity may not be exerted in the case of an essentially chronic, essentially diathetic disease, impregnating the organism so profoundly as to have the twofold property of affecting the whole system, and of developing its manifestations at all periods and all intervals, up to thirty, forty, and fifty years after its origin! We are to regard a disease of this character as inactive, hereditarily, from the father to the child! If this were the case, it would be, in truth, an anomaly most strange, a monstrous departure from all that we know concerning the general laws of heredity.

The syphilis of the father is not inoffensive to the product of conception: this is what theory presupposes; this is what a legitimate induction, based upon the elements of common observation, leads to.

But, in considering a question at once so serious and difficult as this, we should not be satisfied with comparisons, indirect inductions, *a priori* reasonings. It is facts, and precise facts, which we must have. Let us consult, then, clinical observation, and see what it teaches us.

We commence by stating the position as strongly as possible (regretting our inability to make it stronger still) of the partisans of the doctrine that we are about to oppose, and say with them: *Yes, it is true, absolutely true, that we encounter in practice numbers of men, who, having contracted syphilis before their marriage, have be-*

gotten children healthy and exempt from syphilis, their wives themselves remaining healthy and uninfected.

Examples of this kind are observed every day in private practice. MM. Ricord, Cullerier, Notta, Charrier, Durac, Mireur, and many others that I need not mention, have related cases of this character as authentic and convincing as possible. My own personal observation accords fully with that of the authors I have just cited, and I find in my notes (to speak only of the cases coming under my immediate observation) eighty-seven cases in which syphilitic fathers married to healthy women, who remained healthy, have had healthy children, absolutely exempt from every syphilitic manifestation, from every suspicious symptom.*

I am reluctant to relate individual cases here, so frequent and common are they. Nevertheless, there are so many cases observed in which the non-influence of the father's disease upon the child is shown in a manner so manifest, so striking, that you will pardon me for citing some of them, in order to establish your convictions more thoroughly upon this point.

A patient of our distinguished *confrère*, M. Charrier, had been affected with syphilis for several years, when he became the father, exactly at the same date, of two children, viz., one born of his wife, to whom he had communicated syphilis, the other born of a mistress, exempt from every specific antecedent. Now what happened? This: Of these two children, one, the legitimate child, came into the world with syphilis; the other, the natural child, was born healthy, and remained so.†

---

* V. Illustrative Cases, Note 1.

† Charrier, *De l'hérédité syphilitique.*—*Archives générales de médecine*, 1862, t. ii, p. 327.

The natural and irresistible conclusion is, that, when the mother is healthy, the syphilitic influence of the father upon the product of conception is null.

Another example, furnished by M. Mireur : A man marries after eleven months of syphilis, and becomes the father of a fine child, absolutely healthy (the mother remaining still uncontaminated). Now, this child was so healthy, he was so little tainted with the least syphilitic vice, that at the age of two years he contracted syphilis, and from whom ? From his father ! The father had a secondary erosion in his mouth ; he kissed his child upon the mouth, and in this way communicated to him a chancre of the lip.*

But there are two classes of facts more confirmatory still, as they embrace an additional, supernumerary element, viz., the breaking out of syphilitic accidents upon the father, either subsequently to the date of conception or even at the moment of conception. And, nevertheless, children born in such conditions have escaped syphilitic heredity, although the paternal affection was still manifestly persistent, or even exhibited actual symptoms, at the time of procreation. I will explain :

I. It is a matter of frequent occurrence for syphilitic men to beget healthy children, and *afterward* present such and such accidents of syphilis, unequivocal evidences of the persistence of the diathesis at the time when conception takes place.

Example : One of my patients, syphilitic for ten years, married while exempt from every apparent diathetic phenomenon, and became the father of six children. These six children, of whom the eldest is at present eleven years of age, I have had under my observation since

* *Thèse citée,* p. 26.

their birth; I have attended them in all their indispositions, even the slightest, and I am justified in declaring them absolutely healthy. Their mother, moreover, has never presented the least suspicious symptom. Now, *after the birth of the third child,* this man was affected with a tubercular syphilide upon the thorax; and, in addition, *consecutively to the birth of the fifth child,* I again treated him for a gummous tumor of the palate of a very threatening aspect.

Here, then, is a man who has begotten six healthy children, despite a syphilis, active, persistent, and still revealing itself by intense symptoms after the date of conception of these several children.

Again, one of my old patients married, without consulting me, despite a syphilis very insufficiently treated. He has two children, that I have not lost sight of since their birth, and who have always remained free from the slightest specific manifestation. (The elder is at present fourteen years of age, the younger twelve.) Now, this man recently died from cerebral syphilis, not only diagnosticated such from the clinical symptoms, but verified by microscopic examination.

Are not these two cases, and many others which I could add, absolutely convincing? *

But, still, this is not all. I have known syphilitic patients to procreate healthy children, free from the least suspicious symptom, when they were in the *full secondary period,* when they were affected, *even at the moment of conception,* with various syphilitic accidents—when, in

---

* In statistics which will be produced farther on (V. Notes and Illustrative Cases), the reader will find not less than thirty-five cases of this kind, all relating to syphilitic subjects who have engendered healthy children, and who, after the birth of these children, have been again attacked with various specific accidents.

a word, they had not passed that formidable period where the diathesis has, so to speak, its acute crisis, and seems really to be the most pernicious, so far as the dangers of hereditary transmission are concerned.*

I have among my notes cases of this kind, but none of them, certainly, so convincing as a case of the same character which has been obligingly communicated to me by our very distinguished colleague, M. Maurice Raynaud, and which, occurring under conditions altogether special, offering a chronology of quasi-mathematical precision, merits in every respect a place here: A married man contracts syphilis in an extra-conjugal adventure. During several months he invents ingenious pretexts to avoid connection with his wife ; but, finally, one day he forgets himself. The next day he rushes frightened to M. Raynaud, who discovers mucous patches in his mouth. Nine months later, to a day, and without any other subsequent intercourse, the young wife was confined, and confined of a healthy child, which, though now ten years old, has never exhibited the slightest evidence of syphilitic infection.

Thus, here is a syphilitic man who, *the very day when he procreates a child*, presents accidents of the secondary stage, and whose child, nevertheless, is born exempt from syphilis! What more convincing!

You see, then, gentlemen, I dissemble nothing. Very far from it. On the contrary, I insist with all my strength upon the importance of these facts, so curious; of the non-hereditary transmission of syphilis through paternal influence ; for these facts constitute, in my view, one of the most interesting acquisitions of contemporary science ;

* Many cases of this kind are found embodied in the interesting work of M. Notta, to which we have previously made allusion.—*Arch. gén. de méd.*, 1860, t. i.

4

and I need say nothing of the importance they acquire in connection with the subject which we are now considering. The conclusion from the foregoing is, that syphilitic heredity proceeding from the father (and from the father alone, the mother remaining healthy) is much less active, much more restricted, than had heretofore been supposed.

Under given conditions, on the one hand, a syphilitic husband, and on the other a healthy wife, the chances altogether are that the child issuing from this couple will be born exempt from syphilis. This, contrary to the old beliefs, contemporary researches have clearly and positively established ; and this result, certainly, in what concerns us, can not fail to be most consoling.

II. But, this fact recognized, this concession made to the partisans of the doctrine which I oppose, I at once resume my position on the strength of observation, on the strength of clinical facts, and I say to my opponents :

No, it is not true, very unfortunately, in the case of syphilis, that the paternal influence is so immaterial as has been pretended ; still less is it true that it is *null*, that it exercises no influence whatever upon the fœtus. To enunciate such propositions, you must look only upon one side of the question, you must regard only one element of the problem ; for there is a vast difference between the conclusions at which you have arrived and those which are derived from an integral observation of the clinical facts. Judge of this, however, by what follows :

In the first place, if paternal heredity, as we have termed it, exercises its influence in a rare, exceptional manner, still it exercises it *sometimes.* We have seen that children are born syphilitic through the agency of their father, their mother remaining exempt from all contamination. Numbers of cases of this kind have been related by different authors, notably by MM. Ricord, Trousseau, Diday, Depaul, Cazenave, Bazin, Hardy, Bärensprung, Hutchinson, Bassereau, Beyran, Martinez y Sanchez, Liégeois, De Meric, Martin, Parrot, Lancereaux, Kassowitz, Charpentier, Pozzi, Keyfel, Carl Ruge, and others.* I myself have likewise observed some cases, although relatively few, I confess.† I recently met one of my *confrères* who, while doing me the honor to consult me for an old syphilis, said to me that " he had had five syphilitic children, although his wife (examined by him with the most vigilant care), submitted to the most assiduous observation, had never presented the least diathetic symptom."

That, among the facts cited in support of this argu-

---

* See an interesting work of Dr. Léon Richard (*Étude sur l'hérédité dans la syphilis; de l'influence du père.—Thèses de Paris,* 1870), which reproduces *en résumé* a certain number of cases under consideration here.

See also: Piquand, *Influence de la syphilis des générateurs sur la grossesse.— Thèses de Paris,* 1868.

Bricard (Ph.), *De la transmission de la syphilis du père à l'enfant avec immunité de la mère.—Thèses de Paris,* 1871.

Kassowitz, *Die Vererbung der Syphilis.* Vienna, 1876.

Carl Ruge, *Ueber die Fœtus Sanguinolentus.—Zeit für Geburtsh. und Gynäkologie,* B. I.—Analysis in *Revue des sciences médicales,* t. xii, p. 203.

Professor Parrot has recently narrated to me a case of this kind observed by him under special conditions which leave no possibility of error: "A young man, married, with a syphilis in full activity. He had two children, who both presented undoubted symptoms of hereditary syphilis. Now, their mother, closely watched over, minutely examined from time to time since her marriage, has never presented, and still does not present, any suspicious symptom. Without doubt, she remains entirely exempt."

Mr. Hutchinson is still much more pronounced in favor of paternal heredity: "I am firmly of opinion that, in a large majority of instances in English practice, inheritance of syphilis is *from the father,* the mother having never suffered before conception" ("Medical Times and Gazette," December, 1876 ; v. likewise, "A Clinical Memoir on Certain Diseases of the Eye and Ear, consequent on Inherited Syphilis," p. 209, London, 1863 ; "On the Transmission of Syphilis from Parent to Offspring," "British and Foreign Med.-Chir. Review," 1877, vol. lx, p. 455).

† I find in my notes only eight cases of this kind. Even some of these are wanting, I admit, in the guarantees of authenticity which would be required in a matter so delicate and so disputed.

ment, and reproduced in various monographs and reports, there are a certain number to be excluded, to be challenged on account of insufficient guarantees, I do not deny ; rather, I will affirm it, if need be. But will it be possible to challenge *all*, at once and in a lump ? Is it to be believed that all the authors, who have seen, published, and commented upon such facts, have together fallen into the same error, in all failing to recognize syphilis in the mothers of the children that they had under observation ? No, indeed. That is not admissible. However zealous a partisan one may be of the new doctrine, one is not authorized, it seems to me, to throw overboard the whole bundle of former observations which contravene this doctrine, except after a long, very long and ultra-sufficient experience. Now, an experience of this kind is still wanting to us ; so that, actually, in the present state of our knowledge, it is absolutely necessary for us to take cognizance of the facts brought forward in favor of paternal heredity, and to admit this : *That, however rare, however exceptional the hereditary transmission of syphilis from the father to the fœtus may appear to be, it may, nevertheless, exert itself in this way in a certain number of cases.* Consequently, in that which concerns us, this is the first danger which it is necessary to calculate, in order to render a verdict upon the fitness for marriage of a patient tainted with syphilis.

III. But even this, as you will see, and as I am anxious to convince you, is only the lesser side of the question. For, in a manner much more frequent and much less contestable, the syphilis of the father creates for the child other dangers of a graver character. What, then, are these dangers ? To summarize, they consist in this :

1. *Inaptitude for life*, revealing itself by the early

death of the fœtus, either *in utero* or very shortly after birth.

2. Constitutional vices, morbid aptitudes, defects, inherent infirmities, congenital malformations, arrests of development, etc., which I, like many physicians, look upon as constituting the modified, transformed expressions of specific heredity.

We have here, certainly, relative to the question which we are examining, a whole series of considerations well worthy of our careful attention. We are here in the very heart of our subject. Let us insist, then, upon the several points which I have just raised.

I have said, in the first place, that one of the consequences of paternal syphilis for the child may be inaptitude for life, inaptitude revealing itself by *death in utero.* *In other words, a child born of a syphilitic father and of a healthy mother is liable, by the fact of the paternal syphilis, to die before coming into the world.* That is a principal point upon which my conviction is now well established. Formerly, I was struck with the frequency of abortions in families where the husband was infected with syphilis, while the wife remained perfectly healthy. Afterward, I determined to confirm this general impression by instituting a precise inquiry into the matter. With this view, I applied myself to note the results of the union of a syphilitic man with a healthy woman in a very exact manner in all the cases which came under my observation. Now, after several years of investigation in this direction, an abstract of my observations furnishes me with no fewer than fifty abortions occurring under the above-mentioned conditions, and produced without other possible cause to be alleged than the paternal diathesis. And be pleased to note (it is essential to specify this) that

the elements of these statistics have been collected in
private practice, in a *bourgeoise* practice ; that is to say,
in a social medium where the anti-hygienic conditions of
misery, of forced work, of fatigue, of insufficient alimentation, of excess, of debauch, etc., have not played
any rôle as predisposing causes of the abortion. Note
that they have been collected (thus the analysis of
my observations demonstrates) among young women in
very good health for the most part, recently married, presenting no uterine lesion, etc. So that, in all these cases
(excepting two or three at most), no other cause, either
constitutional or accidental, could be alleged as sufficient
reason for the abortion. The abortion remains inexplicable upon the assumption of the influences, predisposing
or determinant, to which it is usually attributable, while,
on the contrary, a common etiological element reunites
all these cases and serves as a common explanation for
them, viz., *the syphilis of the husband.* Is not this well
adapted to enforce conviction ?   ·

Add, moreover, that this fatal influence of paternal
heredity does not manifest itself always by a single abortion. Often it is prolonged ; it is continued in the course
of several pregnancies more or less close together. So
that two, three, four miscarriages sometimes succeed, one
after another, without other explanation than the syphilis
of the husband. The cases of this kind are not rare, I
repeat. I can cite more than twenty examples in my
practice alone.*

Such facts, certainly, are significant of themselves, but
they assume a value much more exact still, when a
counter-proof comes to be added as follows : Apprised by

---

* J. Hutchinson, " On the Transmission of Syphilis from Parent to Offspring."
—" Brit. and For. Med.-Chirurg. Review," 1877, vol. lx.

his physician of the probable cause of these successive abortions, the husband submits himself to a prolonged specific treatment. There occurs afterward a new pregnancy, which results in an infant at full term. Other pregnancies succeed, and these are not less fortunate. Now, then, the evidence is conclusive; how can we deny, under such circumstances, the corrective influence of the treatment upon the syphilitic diathesis, and how can we fail to recognize the influence of this diathesis upon former pregnancies? Successive abortions *before* treatment; fortunate pregnancies *after* treatment; what could be more demonstrative? Now, cases of this kind exist in medical literature. They exist especially in a very much larger number in the memory of practitioners, as I have convinced myself by many conversations with my *confrères.* I myself have observed several, such as the following, for example, which vividly impressed me at the beginning of my career, and which I can not resist the desire to relate to you briefly: Fifteen years ago I encountered, by chance, an old college comrade whom I had long lost from view. We talked together, and he related his troubles to me. "You see me disappointed," he said; "my wife has just had her fourth miscarriage in an early period of pregnancy; and, what is worse, all these miscarriages have taken place without the least cause—without a fall, without imprudence on her part. My wife is large, strong, well developed, perfectly healthy; and, nevertheless, I foresee, to my great grief, that she will never bear me children." The recollection of a certain circumstance then crossed my mind, and I replied: "But, perhaps, your wife is not so responsible as yourself for these successive miscarriages. I knew you many years ago, in the Quartier Latin, with a fine pox, which you did not appear to me to attend to in a very exemplary manner. In your place, I would commence a course of treatment; I would again take mercury and iodide."

Although this interview took place in the street, my advice was followed, and specific treatment was pursued actively. Fifteen months later my friend's wife was delivered, at full term, of a living child, which is now twelve years old. And since that she has had three other pregnancies equally favorable.* The undeniable inference from all this is, that numbers of abortions happening without cause in healthy women admit of no other explanation than that of the husband's syphilis.

The syphilitic influence of the father kills the fœtus in utero. Here is a fact which, supported by observations as authentic as they are numerous, merits a place in science; and I am astonished that it has not been more remarked before. In the second place, this *same inaptitude for life* of the infant procreated by a syphilitic father reveals itself by *death immediately upon, or soon after, delivery.*

In the statistics from which I have borrowed the foregoing facts, I find no fewer than thirty-six other cases of pregnancy (always the issue of a *syphilitic father* and a *healthy woman*) which have resulted at term in infants *born dead* or dying, or sickly, stunted, emaciated, senile children, doomed to an early death. Sometimes, again (a detail curious and essential to know in practice), infants procreated under these circumstances come into the world in a passable or average condition; then, after a few days, after a few weeks at most, they are suddenly extinguished; they die without *disease*, without apparent cause, from one day to another. Of what do they

---

* M. Depaul has related several facts of this kind in his able clinical lectures.

die ? I am unable to say; for, in those cases in which I have had an opportunity to make an autopsy, I have discovered nothing which could explain this kind of death. They always succumb very rapidly, almost instantaneously, and this without an attack, without very pronounced morbid symptoms, to the very great surprise of their parents and of the physician. And, without doubt, they succumb only on account of a congenital vice, of a native debility, of an *" inherent inaptitude for life,"* rather than from a superadded, contingent, incidental morbid cause.

This is not all yet. The more I advance in practice, the more I perceive myself pervaded by the conviction that the influence of a syphilitic father upon his child still reveals itself after birth in various ways : by a general organic debility ; by a constitution enfeebled, impoverished, " delicate," as the common people say, below the normal average ; by a slight power of resistance to morbific causes, which impresses upon incidental maladies a character of pernicious malignity ; by a tendency to nervous accidents, notably to convulsions ; by a tendency to lymphatic and scrofulous affections, etc.

But let me, for the present, reserve this class of considerations until we come to discuss the question of mixed heredity—I mean paternal heredity and maternal heredity combined.

To recapitulate, then, the hereditary influence of paternal syphilis is far from being so innocuous, so slight, so "null," as it has pleased certain authors to assert. They have said that the procreation of an infant by a syphilitic father signified nothing, since the child had nothing to fear from paternal inheritance.* This is a

---

* Criticising this doctrine, M. Voillemier has wittily said : " If one accepts the

great and dangerous error, which good sense reproves *a priori*, and which clinical observation refutes.

In reality, paternal influence, while not exercising itself (as we have previously shown) except in a limited number of cases, is none the less liable to exercise itself sometimes in a manner very positive, very manifest, and then it reveals itself according to three modes :

Either (this is the exceptional case) by the transmission of syphilis to the fœtus ; or (this is much more common) by the death of the child ; or, finally, by inherent degeneration of the germ, which reveals itself subsequently under very diversified morbid forms.

ideas of M. Cullerier, the father would be only the accidental occasion of a child. One would be, in reality, the child of his mother only."—*Gazette des Hôpitaux*, 1854, p. 303.

CHAPTER V.

LET us not lose sight of this other cardinal point : A syphilitic father is dangerous to his children not only in his character of progenitor ; he is, or may become, dangerous to them in his character as the husband of their mother, if I may so express it. In other words, he may endanger them through *the syphilis which he runs the risk of communicating to his wife.* And then, the father and the mother both becoming syphilitic, what must be the fate of the children issuing from this infected couple ?

Ah ! here, gentlemen, is presented a page of pathology distressing to write ; here commences for these families a situation truly heart-rending, which it is necessary to have observed in all its details and in its diverse forms in order to comprehend its miseries.

This situation, which I am anxious to depict faithfully to you in the interest of the grave subject now before us, is here copied from nature in its sad reality :

Two young persons were married a short time ago. The wife has become *enceinte*, and yearns after her title of mother. The two families, full of the sweet hope which preludes the coming of the new-born, impatiently await the result of this pregnancy. Now, what will be the result ? What will happen to the infant procreated under

the conditions which we are now supposing—that is to
say, issuing from a father and mother both syphilitic?
As physicians, we can predict what will happen to it, for,
save rare exceptions, its future is comprised within the
three following alternatives : 1, either this infant *will die
before birth ;* 2, or it *will come into the world with syphilis,* and with all the possible and serious consequences of
infantile syphilis, which, in most cases, is almost equivalent to a sentence of death ; 3, or, finally, it will come
into the world without syphilis, but with a *health compromised, with an innate debility,* and a constitution
impoverished, which will expose it to an early death, with
menacing *morbid aptitudes,* and with a tendency to certain organic vices.

And this is not all, for there may succeed a second, a
third, a fourth pregnancy. It may be that this identical
fate awaits the second, the third, the fourth infant; and
so on until the diathesis has been exhausted by the effect
of time or by the intervention of an energetic treatment.

What a situation ! What affliction for a young couple !
What grief for their two families ! And, in another point
of view, what a social calamity !

That, gentlemen, is what the pox does, or can do, when
the paternal influence and the maternal influence are asso-
ciated ; when both conspire together against the product
of conception.

And these sad results I do not give you as only contingent, as simply possible ; I give them to you, if not as
constant (for, in fact, nothing is constant in heredity), at
least as very frequent, very common, absolutely habitual.

But (for the matter is worth the trouble) let us insist
upon and legitimize the summarized facts which precede :

1st. I said to you a moment ago that the infant born

of a syphilitic father and mother is almost necessarily doomed to one or another of the three alternatives which I have just specified, and which it now remains for us to study in detail.

The first is *death in utero ;* whence abortion, or delivery before term.

Upon this first point there is no possible contradiction. Here medical science is fixed, and securely fixed, by the unanimity of its practitioners.

Open your books ; run over the observations contained in the classic treatises, in the special works, and you will find not only hundreds but thousands of cases, which, in the point of view from which we are speaking, all testify to the same effect, and seem copied one from another. Everywhere and always, it is identically the same observation, stereotyped, so to speak, reproducing itself in the same terms, and summarizing itself thus : " A man in a syphilitic condition marries. In one way or another he infects his wife. She becomes *enceinte,* and aborts in a few months, or is delivered before term of a dead infant."

The intra-uterine death of the fœtus, the offspring of syphilitic parents, is certainly the most habitual expression of the hereditary influence of the diathesis. In truth, this fact is so common, so trite, so accurately verified by numerous observations, that I restrict myself here to enunciating it only. It would be but an abuse of your time for me to stop here in order to cite particular cases.

The pernicious influence of mixed syphilitic heredity, that is, proceeding from both husband and wife, does not always end here in such a situation. Very often, still, this is continued, is prolonged *in a series of successive abortions.*

We have seen many and many cases of unfortunate

syphilitic wives, who have become *enceinte* from contact
with syphilitic men, terminate thus *twice, three times,
four times, five times, six times,* and even *seven times* in
succession, either from abortion or from expulsion before
term of infants dead or moribund.   This very day I can
show you in our wards a case of this kind.   The patient
lying in bed No. 35, St. Thomas's Ward, received syphilis
from her husband several years ago.   Since then, this
woman has become *enceinte six times,* and she has aborted
six times in the third, fourth, or fifth month of her pregnancies.   Likewise, one of my patients, young and well
developed, contracted syphilis from her husband soon
after marriage.   She became *enceinte four times* in three
years, and aborted *four times.*

Cases of this kind have been cited by a number of observers.   But I know nothing of this description comparable to the history of a patient that I treated for a long
time at the Lourcine—a history which you will permit me
to reproduce here briefly.

This woman, large, vigorous, perfectly healthy, married at nineteen years of age.   She commenced by having
three "superb" children, two of whom are still living,
and are, according to her statement, in excellent health.
The third appears to have succumbed to some incidental
disease of an acute form.   After her third confinement,
this woman received syphilis from her husband, which he
had contracted a short time previously in an amorous es-
capade.   Since then, she has been *enceinte seven times.*
Now, what has been the termination of these numerous
pregnancies subsequent to the contagion?   The result is
curious and dismal, in truth:

First pregnancy (after the syphilis): Abortion in the
fifth month.

Second pregnancy : Premature accouchement at seven months and a half. Infant sickly, emaciated, dying the fifteenth day.

Third pregnancy : Accouchement almost at term. Infant born dead.

Fourth pregnancy : Accouchement premature. Infant still-born.

Fifth pregnancy : Accouchement premature. Infant still-born.

Sixth pregnancy : Abortion at three and a half months.

Seventh pregnancy : Abortion at six weeks.

Résumé : Ten pregnancies, of which the three previous to the syphilis resulted in three children at full term and perfectly healthy, and the seven subsequent to the syphilis resulted in four premature deliveries and three abortions ! What fact more instructive? and what could you demand more probatory in support of the thesis which we are developing?*

2d. Second alternative : The infant issuing from a syphilitic couple may be born alive, but it is *born with syphilis,* and bears all the consequences, so grave and so formidable, of hereditary syphilis.

Here, again, long developments are not needed to establish two facts which are patent, which spring, with an evidence unfortunately too manifest, from common, almost daily experience, viz. :

1. That the offspring of syphilitic parents are most ha-

* See Illustrative Cases, Note 11, the relation *in extenso* of this curious case. I owe the communication of another analogous case to Dr. Le Pileur, physician to Saint Lazare. This case is briefly as follows: Wife, syphilitic, becomes *enceinte eleven* times. Of these eleven pregnancies, five are terminated by abortion or by the expulsion of still-born infants at various epochs of gestation. Six others resulted in living children, of whom five died from convulsions, viz.: four the first or second day, and the·fifth at six weeks. One child alone has survived. Eleven pregnancies, ending in a sole case of survival!

bitually born syphilitic, especially in the course of the earlier pregnancies which succeed the infection of the parents,
that is to say, when time and treatment, those two great
correctives of the pox, have not yet exercised upon the
diathesis of the generating couple their attenuating and
depurative influence.   This first fact is neither contestable
nor contested. It is useless, then, to insist further upon it.

2. That children who are born with hereditary syphilis
are exposed, by virtue of this syphilis, to multiple and
most serious dangers. By dint of care, we succeed, indeed,
in curing a certain number of them. But, whatever we
may do, despite all treatment, a very large number succumb. I do not hesitate to confess that my personal statistics of the syphilitic new-born, even when treated, are
truly deplorable as a mortuary table.   Nothing is so murderous as hereditary infantile syphilis. This is a second
fact which again it will suffice simply to enunciate, so commonly is it observed.

3d. Third and last alternative: It is possible that
a child born of syphilitic parents may escape death *in
utero*, or even the syphilis. But it is not yet free from
danger for all that, since the syphilitic influence may still
exert itself upon it in other forms, which it now remains
for me briefly to point out.

I must tell you that we have here to deal with one of
the most difficult and most delicate points in pathology.
Indeed, just as hereditary influences are direct and unmistakable when transmitted from one generation to another
by the reproduction of the same malady, so they become
doubtful and questionable in opposite conditions, that is
to say, when they reveal themselves in the offspring by
symptoms different from those manifested in the parents.
And, nevertheless, *this heredity with dissimilar morbid*

forms, if I may thus express it, is no less authentic than
is the other heredity, with identical morbid forms, only it
oftener escapes attention, as it does scientific demonstration. Such is the case here. The entire profession grants
the syphilitic heredity which reveals itself from one generation to the following generation by symptoms of a
syphilitic order, while they have long discussed, and will
long continue to discuss, the question whether the syphilitic influence of parents can exert itself upon their
descendants by manifestations or morbid tendencies
not directly entering into the list of symptoms of this
diathesis.

As for me, my position is taken upon this question,
which has long engaged my attention, and which I have
studied, I believe I can say, with minute attention. After
having doubted, I doubt no longer, and my present conviction is that the syphilitic influence of parents does not
reveal itself in their children by symptoms of a syphilitic
order only, but also by morbid conditions, by morbid dispositions, nowise syphilitic in themselves, which have
nothing to do with the classic symptomatology of the pox,
which are even as different from it as possible, but which,
nevertheless, do constitute modified expressions of the diathetic state of the ancestors, do constitute, if I may so express it, a sort of *indirect descent* of the pox.

And, moreover, what is there singular, what abnormal,
what inexplicable in this hereditary modality? Does
syphilis have for symptoms in the subject which it affects
only manifestations of a specific order? Is everything
that it produces, everything that it determines in the way
of morbid troubles, always and invariably of a specific
order? Parallel with its peculiar lesions, has it not also
a train of general symptoms? At the same time that it

attests itself in dermatoses, erosions, ulcerations, infiltrations of organs, visceral neoplasms, etc., does it not on the other hand also reveal itself, generally, by phenomena of anæmia, malnutrition, emaciation, impoverishment, and organic deterioration, sometimes also by nervous troubles —in a word, by reactions of a general character upon the various organs? Does not syphilis, as M. Ricord so justly declares, awaken scrofula in the scrofulous? Does it not also awaken *dartre* in the dartrous, as our lamented colleague, M. Bazin,* taught here? Does it not also react upon traumatic lesions, as M. Verneuil and his pupils are now in a fair way to demonstrate? † Syphilis, then, is not simply a disease with syphilitic symptoms. It is a disease of the whole system. It is a disease which creates a general disturbance in the whole organism, which affects, or may affect, that which is commonly called "the health," which awakens, or may awaken, very diverse morbid tendencies—in a word, it is a disease with diversified and polymorphous reactions.

Now, if this is the case, if syphilis is capable of introducing disturbances so profound, and at the same time so complex, into the organism which it affects, what is there astonishing in heredity reflecting these varied morbid dispositions in the product of conception, in the child, the offspring of syphilitic parents?

However, let us leave these theoretical discussions and consider only what observation and the clinic teach us.

The clinic teaches us that children born of syphilitic parents are exposed to certain morbid conditions, to cer-

---

* See *Leçons théoriques et cliniques sur la syphilis et les syphilides,* second edition, Paris, 1866.

† See Henri Petit, *De la syphilis dans ses rapports avec le traumatisme.—Thèses de Paris,* 1875. The reader will find in this estimable work a very complete history of the subject.

tain morbid aptitudes which are produced in them with a significant frequency.

Let us proceed to precise facts.

These children are very frequently remarkable, almost recognizable, I may say, by their *native debility*. They come into the world small, singularly weak and puny, poorly developed, wrinkled and shriveled, stunted, with the "old man look," as it is usually termed. One would call them old people in miniature, with a skin too large for them over certain points. Sometimes, again (a particular sign to which I call your attention), they present on the anterior surfaces of the legs a condition of sub-œdematous puffiness of the integument, which no longer glides over the subjacent parts, but which seems to be united to the cellular tissue and the aponeurotic tissues of this region. Nothing else, however, of a special character attests a well-pronounced syphilitic state in these children, these little old people, as they are called; nothing else indicates the existence of any other malady. And, nevertheless, at the first glance, one judges correctly that they will not live. Even the nurses do not make a mistake in this respect. I have known many to refuse such nurslings, because, they said, "they would not succeed in raising them." Scarcely, in fact, have these children the strength to nurse; "they do not draw," their mothers or their nurses repeat to you; they sleep upon the breast. Then they become more and more feeble, and soon your first previsions are confirmed. These children do not die, properly speaking; they fade out rather than die; they cease to live, for the sole reason that they are not viable, that they are unequal to life on account of the functional insufficiency of their organs.

2. At other times (and here I am going to reproduce a

pathological fact which I have alluded to in a preceding chapter)—at other times, I say, these children present more favorable appearances. They come into the world feebly constituted, without doubt, but, on the whole, with an average or passable development, which permits us to consider them viable. We have reason to hope that with care and a good nurse they will "pull through," as is the case with so many other new-born, who, feeble, puny, and delicate at first, soon develop and become strong in the course of the first weeks. And, in reality, these children continue to live without accidents and without apparent disease. Then, after some days, after some weeks, suddenly they commence to pine away, and rapidly fade out without an attack, without apparent reason, without any morbid incident superadded. Sometimes, even, as I have already remarked to you, they die in a moment, in a manner the most unexpected, the most unlooked for, without their parents and physicians knowing how and wherefore this *sudden death* is produced. I have, now among my notes more than half a score of cases of this kind, and, as an example, you will permit me to cite the following, which I observed with the coöperation of one of our most distinguished and well-known accoucheurs :

A young man contracts syphilis and is not treated for it, or is treated in an ephemeral fashion, altogether insufficient. Some time afterward he marries. His wife becomes pregnant almost immediately. During the course of her pregnancy she begins to be affected with the various phenomena of secondary syphilis. She is confined almost at term of an infant of average size, passably developed, and free from every apparent sign of syphilis. Nursed by its mother, attended by my colleague and myself, this child grows regularly for several weeks without presenting the least morbid symptom, syphilitic or otherwise. All appears to go well, at least relatively, when one morning we learn that the child suddenly succumbed during the night. The evening previous it had been examined by my colleague, who had found it in a quite satisfactory condition. One hour before its death, its mother had held it in her arms and changed its diapers "without remarking anything unusual." In brief, death occurred in a manner absolutely sudden and unexpected.

Note well, gentlemen, these cases of *inexplicable sudden deaths* not preceded by any apparent morbid phenomenon. You will certainly encounter them in practice, for they are by no means rare. Many accoucheurs among my colleagues or friends have told me that they, like myself, have observed them, and almost always in "children syphilitic or the issue of syphilitic parents." Here, then, is a fact which I commend to your attention.

3. In other cases, children born of syphilitic fathers and mothers escape both death and syphilis. But they present themselves with a wretched appearance, with a poor and debilitated constitution, with a condition of anæmia, persistent and rebellious to all treatment, with a vital resistance inferior to the normal average. One has a presentiment, at a glance, that they will readily yield to a slight disease, that they are subjects predisposed to what is called the *malignity*, the occult insidiousness, of diseases. And, in reality, they are often carried off by maladies which could have been easily controlled in subjects with a better established health and a more vigorous temperament.

4. Another point of which I am convinced is, that children, sprung from syphilitic ancestors, present a decided predisposition to *affections of the nervous system.*

A very large number, for example, die from *convulsions*. In examining my individual observations, I find no fewer than fifty cases where children born in these conditions, whether syphilitic or not, suddenly died in the course of one or more convulsive attacks. And, on the other hand, numbers of cases of the same character are found signalized in special treatises or in periodical publications.

Again, these same children are powerfully predisposed to *meningitis*. This is a remark which I made long ago, and I have not been the only one to make it.* I should not be surprised if the pretended successes of iodide of potassium in tubercular meningitis (some cases of this kind have been published, as you are aware) were explicable by the specific character of the lesions for which this remedy was administered.

In its acute forms, this meningitis of the children of syphilitic parents is almost invariably fatal. In its mild, progressive forms, it may spare the life, only to end, most generally, in a state of intellectual incapacity, bordering on imbecility or idiocy. You may be sure that many children, backward, *imbecile*, or *idiotic*, are nothing else than the products of syphilitic heredity.

I have under observation at this moment an example

---

* At the moment I was reading the proofs of this volume, chance furnished me a new and deplorable example of this hereditary influence of syphilis in the production of meningitis.

One of our most distinguished *confrères* from the country came to pay me a friendly visit. The conversation turned upon one of our mutual friends, a physician like ourselves. "You remember well," my *confrère* said, "poor Dr. X——, that we have both treated for a grave, persistent syphilis? Well, he has just lost his third child, which succumbed, from *meningitis*, like the first two. He has no doubt—nor have I, for that matter—that these successive meningites, which have carried off all his children, are the remote results of his old diathesis. . . . However," added my *confrère*, "I, for my part, firmly believe in the hereditary influence of syphilis as a cause of meningitis among infants. I have seen too many cases in my practice not to be convinced upon this subject."

. of the kind, which is too complete and too demonstrative for me to resist the desire to relate it to you. A child is born of a syphilitic father and mother, who have already engendered two syphilitic children, both soon dying. From the first, it does not develop physically ; its growth is retarded ; so that, at twelve years of age, you would take it for a child of six years at most. Toward his thirteenth year he becomes unintelligent, obtuse, as if foolish ; he forgets the little that he knew ; he loses his memory ; he can scarcely find words to speak. He falls into a sort of torpor. Then occurs an acute crisis of encephalo-meningitis, vomitings, obstinate constipation, strabismus, delirium, partial convulsions, tremors, epileptiform attacks, alternating with long periods of resolution and of coma, paralyses, contractures, etc. Specific medication (iodide of potassium and mercurial frictions), although administered very tardily, dissipates all these morbid symptoms with a significant rapidity. But his intelligence is not reëstablished. Far from this, it remains abolished, extinguished, annihilated in every sense of the word ; so completely, that the child is to-day no more, to speak definitely, than a veritable *idiot.**

It is, to my mind, no less evident that the hereditary syphilitic influence (even limited to the father alone) constitutes a predisposition to *hydrocephalus.* This is attested by a number of facts which I have had occasion to record in my practice. I could cite, among others, the

* I have from my colleague and friend, Dr. Tarnier, a case of *congenital idiocy* in a child born from a syphilitic father. "From the earliest period of its life," says this learned accoucheur, "the strange aspect and the general condition of the child had directed my attention toward the search of a syphilitic etiology, although nothing special justified this suspicion. I interrogated the father with this view, and was informed by him that he had contracted syphilis but a short time before his marriage, and had been treated for it in a very insufficient manner only." And also other similar cases, which I could produce.

case of one of my patients, who, having had the imprudence to contract a marriage despite a syphilis not treated,
has had three hydrocephalic children in succession. I
should add that, in my investigation of this subject, I have
encountered here and there, scattered through medical
literature, numerous observations of the same character.

5. Finally, there rises the question of *lymphatism* and
scrofula, which, by certain authors, are regarded as only
disguised forms of hereditary syphilis.

Assuredly, it would be a great exaggeration to regard
scrofula as a degeneration of syphilis. Assuredly, it would
be a serious error, from a pathological point of view, to
make it subordinate to syphilis, to consider it in the light
of a bastard, transformed, metamorphosed syphilitic affection. Scrofula, unquestionably, has no need of syphilis
in order to exist. It exists by itself alone, or, at least, it
is the effect of causes which have nothing to do with the
syphilitic virus. Ordinarily, we encounter scrofulous
children descended from parents who have never presented
the least syphilitic symptom.

But, on the other hand, it is no less certain that syphilis constitutes, if you will permit me the expression, *one
of the affluents* of scrofula. It brings its contingent to
scrofula, by virtue of its being a debilitating, anæmiating
disease, a disease impoverishing the organism, deteriorat-
ing the constitution, ruining the vital forces. It beckons
scrofula in its train, it predisposes to it in the same manner as do all depressing causes, in the same manner as
misery, insufficient alimentation, captivity, etc. And this
action which it exercises upon the health of the parents
is reflected and revealed afterward in the child, by manifestations peculiar to lymphatism in general, and to the
highest degree of lymphatism, that is, scrofula.

Such are, to speak only of facts well established, the states or morbid aptitudes which may be derived from syphilis as hereditary consequences. Still, I am far from saying to you all that I think. For I strongly suspect that syphilis serves as the origin of other functional or organic disorders, such, for example, as congenital malformations, arrests, retardations or deviations of development, spinal curvatures, deafness, keratitis, strabismus, etc. But I pass on from these several points, which might become a matter of dispute, and of which I should not have the right to speak to you with a sufficient degree of certainty.

CHAPTER VI.

MATERNAL HEREDITY.

FROM what precedes, there follows this general conclusion : Hereditary influence becomes veritably *disastrous* when both father and mother are diseased.

That stated, may we now go farther ? Is it possible for us to distinguish in this mixed influence that which is due to the father and that which is due to the mother, that is to say, to estimate the quotum, if I may so express it, of the hereditary reaction of each of these two parents upon the fœtus ? This is a problem more than difficult, and one which it would be impossible to solve in the present state of our knowledge, for the numerical data which would enable us to institute a parallel between the results of paternal heredity and maternal heredity exercised separately are wanting. All that we can say in a general way, avoiding a more minute analysis, is : The syphilitic influence derived from the father reacts upon the child in only a limited number of cases, while the syphilitic influence derived from the mother is exercised upon the child in a manner much more frequent, much more active, and altogether much more dangerous. A child born of a syphilitic father and a healthy mother has numerous chances of escaping both death and syphilis, and the indirect consequences of syphilis.

On the contrary, when a child is born of a syphilitic

mother, the father being free from syphilis, it has but a slight chance of escaping the hereditary influence, in whatever form it may be exercised. One may even predict that it will inevitably die, if the maternal syphilis is of recent date, or if it has not been repressed by specific treatment.

It may be said, very positively, and without any exaggeration, that the syphilitic influence of the mother is veritably *pernicious* for the fœtus.*

The following statistics, collected from different sources, and which I intentionally give separately, go to establish this with a numerical evidence unfortunately too complete.†

I. The first relates to syphilitic women observed in the city, in private practice. It comprises eighty-five cases of pregnancy, which, considered only in their result the most direct and the least subject to error, viz., the death or the survival of the infant, have furnished me with the following figures :

Cases of survival...................................... 27
Cases of death (abortions, premature accouchements, infants
    still-born, infants dead within a short time after delivery) 58
                                                  ——
        Total...................................... 85

Thus, in eighty-five births, fifty-eight deaths, that is to say, in round numbers, *more than two cases of death to every three births.*

A lamentable proportion, to be sure, but much smaller, nevertheless, than the following.

* In order to appreciate the isolated influence of maternal syphilis upon the fœtus in an absolutely rigorous manner, it will be necessary to consider the cases where the mother alone is syphilitic, while conversely the father is healthy. Now, the cases of this kind (especially those which are free from every chance of error) are very rare in practice, and I have not yet succeeded in collecting more than a small number. We are, then, forced to confine ourselves to a parallel between the cases in which the mother is healthy and those in which she is infected. With the first we are already familiar, from what precedes, and we are now about to see what pertains to the second.

† V. Illustrative Cases, Note III.

II. Our second statistics have been recorded of patients observed in the hospital, for the most part at the Lourcine, some at the Saint Louis.

Let us explain in advance, in extenuation of the unfortunate results we are about to present, that, in the patients of this second series, the syphilitic influence was manifestly complicated by other factors, which it would be unjust to ignore, and which are eminently prejudicial to the success of pregnancy, such, for example, as poverty, privations, irregular and insufficient alimentation, excessive labor, fatiguing vigils, debauch, and often professional debauch (the word is strictly exact), excess of every kind, alcoholism, lack of common hygiene and special treatment, etc. Under such conditions, it is evident that the mortality of the children is destined to be increased. This is, in effect, what occurs, but in proportions assuredly much larger than one would suppose.

The abstract of my hospital notes gives the following results in 167 cases of pregnancy coincident with syphilis:

Cases of survival of infant............................. 22

Cases of death of infant (abortions, accouchements premature,
    still-born, infants dead a short time after accouchement).. 145

Total....................................... 167

145 deaths among 167 births, that is to say, in round numbers, *only one infant surviving in seven to eight births!* What a monstrous proportion! What frightful mortality! In truth, this would not be credited, and I myself would not believe it if I had not under my eyes the irrefutable data which have furnished me the elements of this calculation.*

---

* It will not be useless, I think, to add certain commentaries upon these last statistics, the truly frightful results of which demand an explanation. In the first place, the larger proportion of the cases upon which the observations are based

I have not been the only one, moreover, to establish the foregoing unfortunate results. In this same theatre of observation, the Lourcine, Dr. Coffin arrives at figures much more dismal still. Thus, in 28 pregnancies of syphilitic women which were terminated at the hospital, he has verified this:

Infants dead (abortions, accouchements before term,
    death from 1st to 45th day)................ 27 cases.
Infants surviving............................. 1 case alone.

Only one child surviving in 28 pregnancies! What a proportion! *

One of my former pupils, Dr. Le Pileur, at present physician of Saint Lazare, at my request has kindly examined the administrative registers of Lourcine for a period of ten years, and prepared statistics of mortality among children, the issue of syphilitic mothers. This long work has given the following results:

1. In 414 pregnancies, 154 were terminated either by abortion or by the expulsion of still-born infants at different periods of gestation.

have been collected at the Lourcine, that is to say, in a public hospital especially for females, composed in great part of prostitutes, making a business of debauch, and addicted to all excesses, etc. In the second place, I ought to remark that almost all the patients who figure in these statistics were women affected with secondary syphilis more or less recent. Consequently, they were in that stage of the diathesis which is the most pernicious for the fœtus. Let it be added that the great majority of them had never followed any treatment, at least, any systematic treatment, before their entrance into the hospital. And more, we know by experience how the patients at the Lourcine behave themselves—contriving all manner of ruses in order to escape a mercurial treatment, quitting the hospital when scarcely cured of the more visible accidents, only to reënter, and again leave; in addition, observing no regimen, no medication, no hygiene, etc. So that, without fear of departing from the truth, one might regard the preceding statistics as constituted by cases of *syphilis not treated*, abandoned to its own evolution, and exercising upon the product of conception the fullness of its destructive influence.

    * *Étude clinical pour servir à l'histoire de l'influence de la syphilis, du traitement mercuriel, et des ulcérations du col sur la grossesse.—Thèses de Paris*, 1851.

2. Of 260 infants born at term and living, 141 died within a very short time (only 22 survived more than one month).

Let us add up. This makes a total of 295 deaths in 414 pregnancies, that is to say, in round numbers, *almost three deaths in every four births.*

And note again that, among these children considered here as "surviving," there are assuredly a certain number who must have succumbed later directly from their disease.*

Likewise, again, M. Durac, observing at Toulouse, has seen, in 40 pregnancies of syphilitic women, 36 terminate fatally to the infant.†

After such statistics, all commentary would be superfluous. It is only too evident from these figures that the infection of the mother exerts, or may exert, upon the infant an influence the most active, the most noxious, the most murderous. So, then, as regards our present subject, the worst danger which an infant to be born from the union of a syphilitic man with a healthy woman can incur is for this woman to contract the infection from her husband, since, in this new situation, the health and life of the infant will be found most seriously compromised.

And thus, gentlemen, you see how and in what different ways a man with a syphilis not eradicated, contracting marriage, may become dangerous to his children.

---

* That for two reasons: 1st, because hereditary syphilis most often only makes its invasion some weeks after birth, that is to say, at a period when the mother and the infant may have already quitted the hospital; 2d, because numbers of patients of the Lourcine insist upon demanding their discharge when they see that their child is about to die, "not wishing," say they, "that it should die at Lourcine "— not wishing, in reality, that the disease of the said child should testify to their presence in that hospital.

† *De l'hérédité de la syphilis.—Thèses de Montpelier,* 1866.

CHAPTER VII.

Third Point.—A man, who enters into marriage, with a syphilis not extinct, may become dangerous through himself to the interests of his family.

In other words, he may become dangerous to his family by reason even of the personal dangers to which he remains exposed from his disease, from his persistent diathesis.

With this third point, generally neglected, forgotten, sacrificed — I can not explain the reason why — we are about to touch upon the most delicate and the most difficult side of the problem which we are now considering. Here there are no longer questions of pure pathology alone that we have to examine and discuss. *Morality* is about to join itself and enter in line. Reassure yourselves, however. I know to whom I speak, and I will not waste my time nor yours in preaching to the convinced. I shall only need in this new path to invoke certain principles, certain obligations, certain duties which exist inherent in the breast of every honest man, which are unquestioned as they are unquestionable ; and, at the proper time, I will apply them to our subject in strict measure, where they will be indispensable to it.

Speaking to physicians, I have not to remind you, by

way of premise, that syphilis is a serious, a very serious disease, liable to end, when left to itself or insufficiently treated, either in important affections or in serious infirmities ; or even (frequently, much more frequently than is stated or seems to be believed) in a termination more lamentable still—in death. This is a matter of common notoriety, almost hackneyed. But what I wish to particularize, because it directly concerns our present subject, is that, save exceptions rare and of a special order, syphilis but seldom ends in serious or fatal accidents until after a *remote maturity*, that is to say, after a long series of years, for example, after ten, fifteen, twenty years, and more. It is, as you know, in the tertiary period, a period almost indefinite in duration, that the grave manifestations, the veritable catastrophes of the pox, occur. That is to say, syphilis, ordinarily contracted in the foolish years of youth, during single life, has its grave complications only in mature life, when the former gay youth is transformed into a serious man, is metamorphosed into a husband and father. Such are the pathological aspects of the case, are they not?

Now, if this be so, remark then, I beg you, what becomes the situation of a man who, with a syphilis contracted in his youth and not sufficiently treated, presents himself for marriage under such conditions?

The situation, medically, is that of a man who has every chance to be exposed, in a more or less distant future, to the assaults, more or less formidable, of the diathesis. The situation is that "*d'un malade pour l'avenir,*" if I may thus express it, that of a man with health compromised, of a man damaged physically, indebted to the pox, and destined, sooner or later, to discharge that debt.

In such conditions, is it admissible that this man should

aspire to marriage? Is it honest, is it *moral* that this "future sick man" should think of becoming a husband and a father? And, if he consults us, as physicians, to know whether he is fit for marriage, can we, ought we to allow him to engage in this undertaking upon our own responsibility? No; it is not admissible; it is not honorable; it is not moral for a syphilitic subject to contract a marriage in the conditions we have just specified. And, when he comes to ask our advice, our duty is to enlighten him upon this subject, to refuse him the authorization, the free patent—permit the word—which he comes to claim from us, and to explain to him this refusal for the reasons which we are about to indicate.

What, then, is marriage in its completeness, gentlemen? Marriage is not only an affair of sentiment, of passion, of convenience, and of interests. To consider it from a stand-point more practical, and at the same time more elevated, marriage is an association freely entered into, where each contracting party is pledged to bring in good faith a share of health and physical vigor, with the view of co-operating, on the one hand, for the material prosperity of the family, and, on the other hand, for the raising of children—the supreme and sacred end of every union.

Now, what in this case, I ask of you, will be the share contributed to the partnership by a husband, syphilitic, and not cured of his syphilis. His share will be that of a health compromised, hypothecated, burdened with a debt (I again use the word designedly) hereafter due the pox, that pitiless creditor.

On account of the pox, it may happen that this man may experience one day or another such and such serious affections which will ruin his health; such and such an infirmity which will render him incapable of work, inca-

6

pable of earning his daily bread. And then what will become of the family of which this man is the recognized support? What will become of his wife? What will become of his children? On account of the pox, also, this man may die. What may happen, he being dead, to this wife and to these children? Is it admissible, then, that a man should think of creating for himself a family when he is liable to fail this family? Is it admissible, is it right, is it moral that a man should dream of having a wife and children when he offers the possible prospect of widowhood to this wife, of orphanage to these children, of poverty to this family? No, a hundred times no! Also, and I do not hesitate to say it, the man who, syphilitic and not cured of his syphilis, fears not, nevertheless, to append his signature to the marriage contract, commits at this moment a base act, an act immoral and corrupt, an act which good people will be unanimous in condemning.

A comparison will confirm my idea by embodying it in a common illustration : Two individuals associate their interests, let us suppose, in some business, whatever it may be. One contributes his partnership share in good money or in good values ; the other, without the knowledge of the first, furnishes his share in values, doubtful, hypothecated, adulterated, stamped with an inevitable depreciation of market value in the future. What think you of the action committed by the latter? Well, the last is our syphilitic, who brings into the partnership of marriage a depreciated health, a health of poor quality, if I may thus express it, with the prospect, certain or probable, of pathological catastrophes, compromising, or capable of compromising at a given time, the material interests of the association.

In the two cases, the form of the arrangement is very different, assuredly, but the principle remains the same, and there is the same immorality in both. And do not accuse me here, gentlemen, of exaggeration. Do not think that, for the requirements of my cause, I designedly emphasize the situation and darken the picture. It is not so. I speak after what I have seen, exclusively, and without fantastic additions. Unfortunately, it is only too true that, even from the sole point of view from which we are speaking, even, from the sole point of view of the *personal* dangers of the husband, the pox is a frequent source of social miseries the most lamentable, of domestic dramas the most heart-rending. If you doubt it, I have the wherewithal to convince you. I open my notes and I copy from life : A young man marries some years after a syphilis very negligently treated. Six months after his marriage he is seized with cerebral accidents of a specific nature. He dies, leaving a wife and a young child in absolute destitution.

An artist formerly very well known, and quite celebrated on the stage, marries, despite a syphilis which had never been otherwise treated (the expression is his) than "by contempt." He has the good fortune not to infect his wife, and to have a healthy child. But, some years later, he begins to be affected with a tuberculo-ulcerative syphilide, which, still treated with the same stupid indifference, takes on a phagedenic character, plows up the whole face, then destroys entirely the nose and the upper lip, then penetrates into the nasal fossæ, and devours the whole internal bony structure of this cavity, the entire palate, the soft palate, the pharynx, etc. This unfortunate man thus becomes a hideous and infected monster, an object of horror and disgust to all who approach him.

He drags along thus many years in a condition more and more frightful, before ending in a death that to him came too slowly. What a situation! What a spectacle for a young wife, for a child, for a family!—without speaking of moral punishment, and of pecuniary ruin.

Another artist, this one a painter, full of talent and of promise, marries, with a syphilis very insufficiently treated. All goes well during several years. The pictures sell, the little household prospers, and is enriched with a child. Then the husband has an inflammation of the eyes, the nature of which is at first misapprehended, and which, attacked too tardily by specific medication, terminates in complete blindness. Consequence : family ruined, falling into absolute indigence, and forced to inscribe themselves at the bureau of charity in order not to perish of hunger.

A young man comes to consult me for various accidents resulting from a neglected syphilis. I treat him, and all disappears. Some months later, in spite of all my advice and remonstrances, he marries. Twelve days after his marriage, on his wedding journey, he is seized with a violent epileptic attack, the first symptom of a cerebral syphilis, which is soon emphasized by troubles of intelligence, and left hemiplegia. Notwithstanding all my care, he succumbs some months later, leaving his young wife *enceinte.*

A student of medicine acquires syphilis, and judges it proper to treat himself exclusively with the iodide of potassium, not being willing to take mercury. A short time after his doctorate he marries. Some years later he is affected with a slight paraplegia, which is referred to syphilis by common consent of all the physicians whom he consults. Notwithstanding, he still treats himself in a very irregular fashion, "by fits and starts," using his own expression. Finally, he becomes absolutely paralyzed in the legs, and I find him, when he presents himself to me, in a state of absolute incurability. Judge of the situation of our unfortunate *confrère* when you learn that, without resources, he remains infirm, with the charge of a decrepit mother, a wife, and two young children !

A young business man contracts syphilis, and is treated quite regularly for some months. Relieved from all apparent manifestations of the disease, he believes himself out of danger and discontinues all treatment. Three years later, without consulting a physician, he marries. Scarcely married, he communicates syphilis to his wife by a relapse of secondary accidents which occur on the penis. Then he is attacked with symptoms of cerebral syphilis, which I succeed in subduing at first, but which make a new invasion and rapidly carry off the patient.

Epilogue.—The young wife, becoming *enceinte* at the beginning of her marriage, brings forth a syphilitic infant which an active medication succeeds in saving. Very soon she presents multiple symptoms of malignant syphilis— confluent eruptions, cephalalgia, violent neuralgias, ecthymatous eruptions with phagedenic tendency, reproducing themselves when scarcely cured, and ending in covering the body with monstrous sores. Under the influence of such symptoms, her health is altered : emaciation, decline of strength, loss of appetite, digestive troubles, diarrhœa, finally pulmonary tuberculosis, and death from the cachexia—an orphan and without resources, the child has to be relieved by public charity.

A last example—for I could not finish them if I chose to recount all the miseries of this kind which I have witnessed.*

* At the moment that I write these lines, a new and very sad example of the

A manufacturer marries, notwithstanding a syphilis very negligently treated. Thanks to his knowledge of business and the rich dowry of his wife, he founds a great manufactory, which prospers marvelously. Some years later he is affected with gummy periostoses and exostoses of the cranium. There come on gradually cerebral manifestations of various forms : intellectual troubles, vertigo, epileptiform attacks, hemiplegia. He then compromises his fortune and his commercial honor in grandiose and adventurous operations which he is no longer capable of directing, or, to speak more correctly, he would not have undertaken in a sound state of his reasoning faculties, and he is ruined. Finally, he falls into dementia and dies, leaving his wife and four young children in a destitute condition.

What is to be said, gentlemen, of such things, such social calamities ? And what is to be said of those who have caused them, who, after all, are the responsible authors ? In their extenuation, let us admit that they have been more ignorant, more imprudent, than culpable ; let us admit (for this is true in the great majority of cases) that they were not conscious of the injury they might occasion others, of the misery, of the disasters which they ran the risk of spreading around them. But their victims are, on this account, none the less sad examples of the terrible

same order has just presented itself to me. Summoned within a few days to a consultation in a lunatic asylum, I found there a young man affected with a grave cerebral syphilis, and a prey to the most violent delirium. His condition was such as to leave little ground for hope. Now, the history of this patient is traced in that of all the subjects that we have just been considering. At nineteen years he contracted the pox, and was treated for it only just long enough to dissipate the apparent manifestations. Later, he married (about sixteen months since), notwithstanding his syphilitic antecedents. He became a father about a month ago. Being without means, he was dependent upon his labor. What a situation for his wife, what a future for his child!

consequences which may result from their indifference, their recklessness, their thoughtlessness. Well, at any rate, let not these lamentable examples be lost; let them serve as lessons to show us our professional duty, or, rather, the *social* duty which devolves upon us under such circumstances. And this duty, which you have anticipated, is this:

If it is not the province of men of the world and of patients to know what the effects of the pox not treated may be, after a long interval, it is our province to know this and to instruct those who are ignorant. It is our mission to enlighten upon this point patients who come to us, and, more especially still, those who come to consult us upon the propriety of marriage, notwithstanding a syphilis insufficiently treated, and which remains menacing for the future. It is our mission to divert from marriage every patient who presents himself to us under such conditions, to dissuade him from it to his own great advantage and to the great advantage of others, to show him the abyss about to open under his feet, to reveal to him the dangers to which he would expose his future family by a premature union, and to say to him, finally, this, with the authority of our science and our character: " No, sir, no; it is not allowable, in your present condition, for you to dream of marriage, considering only the personal risks to which you remain exposed from your old disease. Until now you have thought it best to live with the pox, to preserve the pox. That was your right, and no one had anything to say, for you were single and, consequently, alone liable for your imprudence. But now, since you aspire to marriage, the situation becomes very different. To marry is to have charge of lives; and, since you do me the honor to consult me, you make it my duty to remind you that you have not the moral right to associate others in your personal risks—that is to say, to make a wife and children share the possible consequences of your disease."

Here terminates, gentlemen, the first part of this exposition.

I have told you how a syphilitic man may be, or may become, dangerous in marriage. I have endeavored to show you that he may be dangerous in a triple manner: to his wife, in transmitting to her the disease with which he is affected; to his children, by way of heredity; to his family, from the personal risks to which he remains exposed.

This will serve us as a point of departure and a basis for the discussion which we are now about to open. From what precedes, the natural conclusion is:

1. That marriage should be forbidden to every man who still presents a syphilis sufficiently active to be dangerous.

2. That, conversely, it may be permitted to every man in the opposite conditions.

But such general facts are not sufficient for the solution of the essentially practical problem which we are discussing. It is necessary that we now grasp the question more closely, that we descend to details, and search out the clinical elements from which we may be able to judge whether a syphilitic subject has or has not ceased to be dangerous in marriage, and whether we should accord him the authorization which he comes to claim from us, or flatly place our veto upon his projects of a union.

CHAPTER VIII.

CONDITIONS OF ADMISSIBILITY TO MARRIAGE — ABSENCE
OF EXISTING SPECIFIC ACCIDENTS—ADVANCED AGE OF
THE DIATHESIS.

GENTLEMEN : The natural order of our subject now
leads me to the discussion of the following question :

In what conditions does a patient affected with syphilis
cease to be dangerous in marriage ? or, what amounts to
the same thing, in what conditions does he become *admissible to marriage*, if I may thus express it ?

Now, as we have been hitherto unrestrained in our
progress toward determining the perils which affect a
syphilitic man in marriage, and in stating theoretically the
general principles of the admissibility or of the non-admissibility to marriage from a medical point of view, so now,
in passing from theory to practice, we are about to encounter embarrassment and difficulties in appreciating the
variable conditions, multiple and complex, of particular
cases.

And this embarrassment, these difficulties, we shall
experience so much more pointedly, as we are now treading upon ground not yet cultivated. The experience of
our seniors, of our predecessors, of those whom, with just
respect, we regard as the masters of our art, is almost
entirely wanting here. And, in fact, gentlemen, you may

run through the classical works, you may examine the special treatises, but you will nowhere find this grave question of the marriage of syphilitics confronted, discussed, debated. Without doubt, you will find here and there certain general hints, certain indications—always more or less vague—incidentally thrown out upon this subject. But nowhere, I assure you from my own experience, will you encounter a veritable programme formulated *in extenso*, or even outlined, upon this matter. All remains to be done, or very nearly so, and this constitutes by no means the least embarrassment of him who has the honor to now address you.*

Let us venture, nevertheless, to attack this difficult and perilous problem, taking for guides, on the one hand, the principles which we have established in our preceding lectures, and, on the other hand, the results furnished by the clinic. In my opinion, according to what I myself have seen, and the results of my reading, the principal conditions which a syphilitic subject ought to satisfy in order to have the moral right to aspire to marriage (that which I will term, by abbreviation, *conditions of the admissibility to marriage* of a syphilitic subject) may be summarized in the following programme:

1. ABSENCE OF EXISTING SPECIFIC ACCIDENTS.
2. ADVANCED AGE OF THE DIATHESIS.

---

* It would be a serious injustice, nevertheless, not to mention here with praise the names of two contemporaneous physicians who have specially treated some of the questions bearing upon our present study, viz., M. Edmond Langlebert, author of an interesting work, very admirably written, upon *La syphilis dans ses relations avec le mariage* (Paris, 1873). The reader will there find several chapters exhaustively studied, and stamped with great clinical judgment. Unfortunately, it is to be regretted that the author should have allowed himself to be diverted from his principal subject by devoting a considerable portion of his book to foreign questions. And also M. Diday, who, in several of his publications, notably in his *Thérapeutique des affections vénériennes* (Paris, 1876), has treated the same subject with that sparkling verve, that vivacity, that humor, which every one recognizes as his.

3. A CERTAIN PERIOD OF ABSOLUTE IMMUNITY CONSECUTIVE TO THE LAST SPECIFIC MANIFESTATIONS.

4. NON-THREATENING CHARACTER OF THE DISEASE.

5. SUFFICIENT SPECIFIC TREATMENT.

Such is, at least according to the results of my experience, the *ensemble* of conditions medically to be required of every syphilitic patient in order that the doors of marriage may be opened to him. If the said patient satisfies all these combined conditions, I consider him fit to become a husband and father without danger.

In the contrary case, I do not consider myself authorized to grant my consent—the moral authorization which he comes to ask of me. I dissuade him from marriage ; I forbid his marriage with all my power. But let us enter into details. Let us explain, comment upon, and justify this programme, which I am very far from presenting to you as definitive, as not susceptible of modifications, and of further amendments, but which seems to me at least to contain the principal conditions to which it is necessary to restrict every syphilitic subject aspiring to marriage.

First condition : ABSENCE OF EXISTING SPECIFIC ACCIDENTS.

This first point certainly will not arouse dispute. It is elementary, in fact, for all the world—for the people as well as for physicians—that the first obligation for a syphilitic candidate for marriage to fulfill is, that he present no syphilitic accidents at the very time of his marriage. For the existence of the least syphilitic accident is a flaming evidence of the disease, of the disease not only potential but actual. And it matters little, moreover, whether this accident be of a transmissible nature or not, for—1. If it be of a transmissible nature, the contraindication of marriage is as express and absolute as possible ; 2. If it be not of a contagious nature, it none the less reveals a permanent diathesis, with all its dangers and consequences. But we need not insist upon this, for the evidence is too direct. And one might be astonished that a proposition like this, the absence of accidents at the epoch of marriage, should even need to be enunciated. *A priori*, in fact, one would scarcely believe that men could be found so destitute of moral sense, so ignoble, so shameless as to dare become husbands with the *existing* accidents of the pox.

And, nevertheless, do not deceive yourselves, gentlemen, this incredible audacity is sometimes encountered. You will find certain cases already signalized in medical literature. For my part, I have witnessed such cases a dozen times. I have seen (and I hold up the fact to public indignation) people marry while presenting *the very day of their nuptials* such or such syphilitic symptoms, as cutaneous syphilides (palmar psoriasis, papulo-squamous syphilide, ecthyma of the legs), mucous patches of the mouth or throat, genital mucous patches, specific sarcocele, accidents premonitory of cerebral syphilis.* I have in my notes the histories of two individuals, who, in spite

---

* The last case to which I make allusion is truly so extraordinary in all respects as to merit special mention. A young man, syphilitic for several years, allowed himself to be engaged in marriage despite various cerebral phenomena, the nature of which, moreover, he but imperfectly comprehended (heaviness of the head, passing vertigo, little aptitude for work, change of character, and, especially, defects of memory). The day of the wedding arrived, but the bridegroom did not appear at the ceremony. He was sought after, and was found occupied with nothing but sitting at the corner of his hearth, reading a paper, *having totally forgotten that he was to be married that day.* Nevertheless, the affair was proceeded with, and (it can scarcely be believed) the marriage took place. The cerebral trouble, be it well understood, became aggravated after this. Some months later, a separation of the married couple had to be pronounced, because of the abuse, cruelties, and violences of the husband toward his wife. Then the patient had an attack of mania, presented various accidents more and more intense of specific encephalopathy, and finally fell into dementia. An English physician of my acquaintance has informed me that he had observed a case almost identical with the preceding.

of my most energetic protestations, were married when they each had on the penis an indurated chancre in the stage of full development. That is not to be believed, is it? But it is true, I assure you, I give you my word, and it proves once more that "the truth is not probable."

What motives, what morbid incitements thus impel certain persons to marry under such conditions, in spite of actual syphilitic symptoms? This is a question, a subject of study, which concerns the philosopher, the moralist, rather than the physician. It is not, however, a matter of indifference to us, for we often need, in the exercise of our vocation, to understand the moral as well as the physical pathology of our patients. Allow me, then, to say a few words in regard to it.

Now, according to my observation, the motives that influence certain persons to a course of action so unjustifiable are not ordinarily those which one would be inclined *a priori* to suppose, viz., ignorance or interest. Without doubt, there are persons who marry in an active stage of the pox from an absolute ignorance of the dangers to which they expose their wives, their future children, and themselves. They do not know what they have, they make no account of it; they have not, from thoughtlessness or stupidity, even dreamed of consulting a physician. These are the simple, the indifferent, the imbecile.

Without doubt, there are others who understand perfectly both what they have and what they may transmit; who take an exact account of the situation, and, while appreciating all its dangers, nevertheless brave these dangers, because they have a superlative interest in braving them, viz., a dowry to secure, a situation to make, a "position" to gain. They are the audacious and the infamous.

But most often this is not the case. In general (at least according to the result of my personal observation) the individuals who are led on to this revolting act of marriage, while the pox is in full activity, are frivolous and of feeble character, who foolishly, thoughtlessly, become engaged to be married when they are as unfit for it as possible ; then, when the fatal moment arrives, they find themselves driven into a corner from which they dare not escape. Although very much ashamed of the act which they are about to commit, while regretting it, deploring it even in their inmost conscience, they have not the courage to recede, to withdraw, for fear of a scandal, of the publicity of a rupture without avowable motives, for fear of what people may say, of public malice coming to suspect their disease.* In brief, to save appearances they commit nothing less than the worst cowardice.

Are these last less culpable than the former ? At any rate, they only arrive at the same result in other ways.†

---

* Example: Some years ago a young man from the country came to ask my advice for certain existing accidents of secondary syphilis (buccal syphilides, alopecia, crusts of the hairy scalp, etc.). The consultation completed, he added, with a confused air, that he was engaged in matrimonial projects, and finished by avowing that he was even contemplating a quite early marriage. I hastened immediately to declare to him that it was absolutely impossible for him to carry out such projects in the condition in which he was, and I gave him the reasons : I insisted energetically, seeing him little inclined to allow himself to be convinced, and I unfolded to him the entire series of dangers to which he was about to expose himself and his future family. Now, to all my arguments this young man opposed an obstinate response, always the same, viz., that "he is constrained to marry for fear that his disease might be suspected." "I should like to follow your counsels, doctor, but this is not possible under the circumstances. What motive can I assign for a rupture that shall satisfy the two families ? What would they say of me in my little village in the country? They will suspect or succeed in finding out the true motive of my withdrawal, and then I shall be ruined, disgraced," etc. And some months later I heard, indirectly, the news of his marriage.

† If I did not fear to exceed the limits of my programme, I would annex to this chapter some considerations relative to those subjects who marry during the actual period of the *incubation of syphilis.* Cases of this kind are, of course, very rare,

Second condition: ADVANCED AGE OF THE DIATHE-SIS.

With this second point we touch upon one of the most important and essential conditions of our programme. In a general manner, indeed, it may be stated as an axiom : *The more recent the syphilis of the husband, the more numerous, the more serious will be the dangers which he introduces into marriage.* Whence this corollary : The older our patient's syphilis, the more shall we be authorized (save for special indications of another kind) to tolerate his marriage.

Let us justify the preceding proposition. In the first place, let us examine the question from the point of view of the dangers of infection for the wife.

Without possible contradiction, it may be asserted that a young syphilis is especially dangerous, in so far as *con-*

but nevertheless worthy of attention. They may be thus summarized : A healthy man, some days before his marriage (a fortnight, for example), has connection with a woman infected with syphilis, and receives the contagion from her. As the first symptoms of the disease are always separated from the date when contagion was effected by a period of three weeks, on an average, sometimes a month, and even more, this man marries with the appearance of perfect immunity. It is not until from eight to ten or fifteen days afterward that the initial phenomena of infection begin to reveal themselves, under the form of one or more local erosions. So that the syphilis contracted before marriage does not reveal itself until after marriage, thanks to the prolonged incubation which always preludes its outbreak. Now, what happens in these conditions ? It is that the husband, not initiated into the secrets of syphilitic incubation, and believing himself exempt from all possibility of contagion, scarcely pays any attention to the lesion which has just commenced to appear upon him. He is far from supposing that this lesion may be of a contagious nature ; he mistakes it for an " abrasion, a scratch, an insignificant trifle." Consequently, he does not abstain from intercourse with his young wife, and in this way he transmits the syphilis to her. I have already observed four cases of this kind ; and in these four cases the contagion was invariably transmitted to the young wife. As an example, I will hereafter relate the history of one of these cases (vide " Notes and Illustrative Cases," Note IV). The possibility of a contamination of this kind in marriage is scarcely known, and it would seem to have but little engaged the attention of clinicians. For the interests of all, it is desirable that it should become of more general notoriety.

tagiousness is concerned. In fact—1. It is a common notion that the disseminated, scattered manifestations of the
diathesis, which, under the name of mucous patches, or,
better, of erosive, papulo-erosive, papulo-ulcerative syphilides, etc., so frequently affect the various mucous surfaces
and the skin—it is a common notion, I say, that these lesions especially belong, chronologically, to the earlier stage
of the disease, to what is termed the *secondary period*. It
is in the first months, in the first two or three years of the
infection, that they are observed almost exclusively. Now,
the contagiousness of such accidents needs no demonstration to-day. We can even say that accidents of this order constitute the principal source which feeds and propagates the pox among us. 2. Every one knows, in the
second place, that, in the first two or three years of the
diathesis, the morbid manifestations of which we have just
spoken are more particularly apt to multiply, to recur,
and that with a persistence sometimes desperate. Let
us cite, for example, the buccal mucous patches which,
among smokers more particularly, are produced and reproduced with many repetitions in the course of the first
months or the first years of the disease. 3. In addition,
at this same period of the diathesis, there are *two centers
of predilection* which influence the morbid determinations,
viz., the *mouth* and the *genital organs*.

Now, it is precisely these two localities which are the
most dangerous from the point of view which now occupies us, for it is from these centers that the contagion will
naturally have the most numerous chances to transmit
itself in marriage.

Let us add still another consideration: secondary
syphilis is particularly dangerous, as regards contagion,
from *the benignity, even, of its accidents*. Very often the

lesions which it determines upon the mucous membranes, in the mouth or on the penis, notably, consist only in very superficial erosions, slight in extent, almost simply desquamative. Now, such lesions may easily pass unperceived, even by careful persons, attentive to their health. They are also apt to be confounded with common, ordinary, insignificant erosions. On the penis, for example, they are frequently taken for simple excoriations, for herpetic or other inflammatory abrasions. In the mouth, they pass not less commonly for aphthæ, fissures, etc., "local irritations from the cigar or cigarette," etc. Altogether, for one reason or another, *one does not mistrust them*, so inoffensive are they in appearance. And this, indeed, constitutes the danger, for such slight, benign accidents do not seem to necessitate continence, and they thus become the origin of frequent contagion in marriage. This is a point which I restrict myself to simply indicating for the moment ; hereafter I shall have occasion to return to it in detail.

Such, gentlemen, are various reasons on account of which a *young* syphilis is so formidable from the point of view of contagion. Quite the contrary, at a more remote period, and *a fortiori*, in an advanced stage of the diathesis, these same dangers of contagion no longer exist, or, at least, only present themselves in a manner much more restricted, relatively, and much less common. And this is on account of reasons exactly the opposite, viz. : because an old syphilis reveals itself only by manifestations infinitely less multiple, more discrete, less subject to recurrence, etc.; because it no longer affects with the same predilection those two centers so favorable to contagion, the mouth and the penis ; because the lesions which it determines in this period consist no longer in minute,

7

superficial erosions, susceptible either of passing unperceived, or of being confounded with inoffensive accidents
of a common order, but in large, profound, important,
persistent ulcerations which could neither escape the attention of the patient, nor permit the possibility of contagion through indifference, inadvertence, etc.

And do not, gentlemen, regard what precedes as pure
theory, but, on the contrary, as conclusions deduced from
experience. In addition, consult the results of the clinic,
and search at what period of the diathesis contagion is
principally produced. Observe, especially, who are the
husbands that communicate syphilis to their wives. I
have made this investigation for my part, I have just
examined my notes on this subject, and without qualifi-
cation I have arrived at this:

If not always, at least in an enormous majority of
cases, the husbands that communicate syphilis to their
wives are those who have entered into marriage with a
syphilis still young—that is to say, with a syphilis dating back several months, one year, two years, more rarely
three or four years.

When a man marries with a recent syphilis still active,
the infection of the wife is an almost constant occurrence.
On the contrary, much more rare are the cases where the
infection of the wife is produced when the husband's
syphilis is more or less old—that is to say, dates back six,
eight, ten years or more. Upon these two points, I repeat it, my observations are precise—as positive and convincing as possible. And I recapitulate by saying: *Syphilitic contagion in marriage is much the more to be feared
for the wife the more recent the date of the syphilis of the
husband.*

II. Likewise, from the point of view of hereditary influence, an advanced age of the paternal syphilis is a condition equally favorable.

It is a fact remarked of ancient date, recorded in a positive manner by various authors, that the syphilitic influence of the father upon his children undergoes a progressive decrease in proportion as the diathesis becomes old. Thus, for my part, in those cases where I have observed the syphilis pass directly from the father to the child without contamination of the mother, I have remarked that the paternal affection was always of recent date—that is to say, it has never exceeded, at the maximum, three or four years. Beyond this term, I have never clearly made out the transmission of syphilis by paternal heredity.

Another proof still is furnished to us by those cases of successive abortions which I have already referred to as the possible consequences of the syphilis of the husband. More than once this has been observed ; a healthy woman, exempt from syphilis, begins by having several *miscarriages ;* then (the first improvement being realized under the sole influence of time, the husband still continuing to neglect treatment) this same woman does not abort any more, she only is confined *before term*, and always of a dead infant ; then, second stage of improvement (since we are compelled to characterize this by the term improvement), she is delivered *at term* of an infant dead or destined to a speedy death ; still later, at length she brings forth one or more *living* infants.

What could more thoroughly demonstrate the normal decrease of the hereditary syphilitic influence under the sole action of *time ?* A *propos* of this, allow me to cite a curious case related by one of our English *confrères*, well known among us, Mr. Hutchinson :

A physician contracts syphilis and treats himself for about six months. Believing himself cured, and free from all apprehension, he marries three or four years later. His wife remains *healthy* and becomes *enceinte eleven* times. Now, gentlemen, look at the results furnished by these different pregnancies, and mark the progressive attenuation which the diathesis undergoes under the sole influence of time :

First pregnancy : child *still-born.*

Second pregnancy : child *still-born.*

Third pregnancy : child born *alive*, but *syphilitic* and *dying*, with the characteristic symptoms of hereditary syphilis.

Fourth pregnancy : child born alive, but *syphilitic* and *dying*, likewise from syphilis.

On the contrary, the last seven pregnancies furnish children who, although syphilitic, resist the disease and all remain living.* And, moreover, could one call in doubt the attenuating and corrective influence of *time* upon paternal heredity when this same influence shows itself in so positive and evident a manner upon maternal heredity, or, collectively, upon mixed heredity derived from the two parents? Is not this a fact absolutely demonstrated, is not this a veritable pathological law, the gradual diminution, then the final extinction of the syphilitic reaction of the parents upon the children? Examples quite convincing in this respect have been given by different authors—by Bertin,† by

---

* Memoir cited, " Brit. and For. Med.-Chirurg. Review," 1877, vol. lx.

† Here is the summary of the curious case of Bertin, to which allusion is made : Father and mother syphilitic. First pregnancy : abortion at six months ; infant still-born. Second pregnancy : abortion at seven months ; infant living eight hours. Third pregnancy : accouchement at seven and a half months, of a dead infant. Fourth pregnancy : accouchement at term ; child syphilitic, surviving

M. Diday,* by M. Bazin, † by M. Roger, ‡ by M. Kassowitz, § and many others. But none is more cogent and better calculated to carry conviction than a very curious case of M. Mireur, summarized thus :

A young mason contracts an indurated chancre, and marries at the very *début* of the secondary stage. He does not fail, for he can not, to infect his young wife immediately.

After this there occur eight pregnancies, the results of which are unfolded according to the characteristic tendency of the disease, the *husband and wife remaining exempt from all treatment.*

Now, these eight pregnancies terminate in the following manner :

First pregnancy : *abortion* at fifth month.

Second pregnancy, *abortion* at seventh month.

Third pregnancy : accouchement before term; *child dead.*

eighteen days. Fifth pregnancy: accouchement at term; child syphilitic, surviving six weeks. Sixth pregnancy: accouchement at term; child syphilitic, but surviving.—*Traité de la maladie vénérienne chez les nouveau-nés,* Paris, 1870, p. 142.

* *Traité de la syphilis des nouveau-nés et des enfants à la mamelle,* Paris, 1854, p. 183.

† *Leçons théorctiques et cliniques sur la syphilis et les syphilides,* Paris, 1866, p. 164.

‡ Summary of an observation related by M. Roger : Father and mother syphilitic. First pregnancy: accouchement at eight and a half months ; infant dead. Second pregnancy: accouchement at term; infant dead. Third pregnancy: Accouchement at term ; child living but syphilitic, covered with lumps when one month old, and dying at four months. Fourth pregnancy : accouchement at term; child living, strong and well nourished. From the second to the third month this child was affected with a syphilide limited to the buttocks, afterward with a very intense coryza ; died at eight months. Fifth pregnancy: accouchement at term ; " child syphilitic, like the others, but in a less degree of severity (roseola simplex). Treated by mercurial medication, it was completely cured. It is now seventeen months old and perfectly well."—*Étude clinique sur la syphilis infantile (Union Médicale,* 1865, t. i, p. 147).

§ *Die Vererbung der Syphilis,* Vienne, 1876.

Fourth and fifth pregnancies: *Children living*, but both *syphilitic*—the first dying at thirty days, the second at one month and a half.

Sixth, seventh, and eighth pregnancies: children *living* and healthy.*

Such a fact is sufficiently eloquent of itself to render all comment unnecessary. Now, if it be thus for mixed heredity, one can scarcely comprehend that it could be otherwise for paternal heredity alone.

Then, time exhausts the paternal syphilitic influence and renders it less and less dangerous for the offspring; here is a point which appears to be well demonstrated.

From this there follows the practical conclusion that with syphilis all the chances are in *waiting the longest possible time* before aspiring to the rôle of father of a family. This, at least, seems to me to result in a manner as direct as possible from the facts observed by me up to this time.

III. From the point of view of the personal risk which a syphilitic subject carries with him into marriage, an advanced age of the diathesis is again a condition certainly favorable, and constitutes an additional guarantee — a guarantee not positive and absolute, be it well understood, but at least relative.

And, in reality, the advanced age of the diathesis permits us, in a certain measure, to better appreciate the "quality" of this diathesis, its degree of intensity, its general prognosis. Not, assuredly, that in the case of syphilis the past is always, as it has been termed, "the

---

* Thesis cited, p. 91. A curious point is that consecutively to the last three pregnancies, which resulted in living and healthy children, the father and mother still presented various syphilitic accidents of a grave form, such as "gummous and ulcerated tubercles, abundantly diffused over the limbs."

mirror of the future"; far from it. Not that a syphilis primarily benign may not result in serious or mortal accidents at a remote period. But it is always the case that, at a period more or less advanced, certain accidents, certain menacing, malignant forms of the disease, are no longer to be feared. Moreover—and this is the essential point — the stage comprised between the début of the disease and the present time may have been employed advantageously in a long and salutary treatment, which confers the best and most substantial of guarantees in all such cases.

Hence, in every respect, you see, the long standing of the disease constitutes an essential, an indispensable condition of admissibility to marriage.

In my opinion, a syphilitic subject has neither the privilege nor the right to aspire to marriage unless his disease dates back a certain time, unless it has already behind it a certain duration. This I positively affirm, and lay down as a principle, it being my absolute conviction, now based upon a very large number of observations.

Now, you will very properly press me further, and, demanding of me more precise indications, will say : "Be it so ; a certain duration of the disease is indispensable for admissibility to marriage. But speak more definitely ; state in precise figures what is, what in your opinion ought to be, this duration ? " Ah ! here comes the delicate point. As long as we have to deal with general terms solutions are easy. But when it becomes necessary to fix an arithmetical measure, when we come to figures, the difficulties begin. Nevertheless, the same facts which have served me in stating the general rules which precede will permit me, within a certain limit, to satisfy you, and I shall reply as follows :

In the first place, the age of a syphilis is not the sole fact from which one may determine the admissibility or non-admissibility to marriage. This decision embraces other factors, involves other important and paramount conditions, of which we shall speak shortly, such as the nature of the anterior specific accidents, the "quality" of the diathesis (you will see what I mean by this word), the intervention of a sufficient treatment, the influence exercised by such treatment, etc. So that the age alone of the disease can never be taken as a basis to solve the problem before us, namely, to decide whether a syphilitic subject has or has not become fit for marriage.

With these reservations, I shall enter upon the question to which you await a response, and at once give you my profession of faith, which may thus be summarized :

To regard for the moment only the fact of the age, I do not think that a syphilitic subject should be permitted to think of marriage until after a *minimum period of three or four years* devoted to a most careful treatment.

Three or four years is, in my opinion, the *minimum* (note well the word, I beg of you), the necessary indispensable minimum, in order that the diathesis may be sufficiently weakened under the double influence of time and treatment ; in order that the patient, restored to ordinary conditions, may properly aspire to become husband, father, head of a family.

Yes, three or four years ; for it is not too much. My requirements are not excessive. *A longer time would be better*, I am quite certain. For, with syphilis, it is always advantageous to wait, to defer, when there are such interests at stake as those of a young wife and an entire family. So that I affirm anew the proposition which I have just formulated, and which is the result of experi-

ence : Short of the *minimum* above cited, there is everything to fear, and the dangers of syphilis in the husband are revealed by catastrophes which, if not inevitable
and constant, are at least frequent and habitual. Beyond
that term, the dangers from the syphilis of the husband
decrease and disappear, if not in a manner absolutely
certain (for mathematical certainty is and ever will be
unattainable in such matters), at least very frequently, so
frequently, at any rate, as to warrant us in allowing marriage. Still it is necessary, be it understood (and here is
a point upon which I do not hesitate to insist again), that,
during the period in question, the curative influence of
a systematic, methodic, active, and prolonged treatment
shall be added to the corrective influence of time.

Before three or four years are past, I could never dare,
for my part, whatever may have been the activity of the
treatment, to grant a clear license of marriage to a syphilitic subject. For I have seen the saddest, the most unfortunate consequences succeed premature unions of this
kind. After three or four years usefully devoted to a
depurative medication, I believe that I am authorized by
experience to *tolerate* marriage, unless there be some particular contraindication noticed in some other portion of
my programme. And this because I have seen numbers
of syphilitic subjects marry in these conditions without
becoming noxious to their wives or their children.

By design, I have just said that in the conditions cited
I considered myself right in *tolerating* marriage. And,
in fact, I tolerate it in such a case rather than counsel it.
I tolerate it because I consider the term of three or four
years as strictly sufficient to protect the interests which I
have at heart. But I will frankly say that a longer delay
would be much more satisfactory to me, as affording more

effective guarantees. In the case of a patient whose syphilis (besides having been well treated) dates back six, eight, or ten years, I feel much more at ease in giving him a clean bill of health ; and this, I repeat again, because, from numerous points of view, the security increases with the age of the diathesis.

So, practically, my rule of conduct is the following: When consulted upon the propriety of marriage by a patient whose syphilis (although regularly treated) dates back only three or four years, I always begin by counseling him to wait, to defer his project of union, and I insist anew upon treatment, with the view of increasing and perfecting his chances of security. But, if, nevertheless, the patient urges serious and paramount reasons for an immediate marriage, and if, in addition, he satisfies all the other requirements of my programme, I do not think that I have the right to thwart his projects. I *tolerate* his marriage under these circumstances. I grant him the medical authorization which he comes to claim of me—not, however, without adding certain advice, certain indispensable recommendations, of which I shall speak hereafter at the close of this exposition.

CHAPTER IX.

*Third condition : Stage of immunity sufficiently pro-
longed after the last specific manifestations.*

A third condition which I regard as indispensable in
this matter is, that a period more or less prolonged between the last manifestations presented by the disease and
the date fixed for marriage should elapse *without specific
accidents.* That is to say, before a patient has the right
to think of marriage, it is necessary that he remain exempt from any diathetic manifestation during a time sufficiently prolonged, and that this lapse of time, which I
designate as the *period of immunity,* pass without any
specific symptom, without any outbreak of the syphilis.

Now, this period of immunity constitutes a necessary,
indispensable guarantee for admissibility to marriage for
several reasons: In the first place, it has a significance.
It proves that the diathesis has passed its acute stage, I
mean that especially formidable period where the syphilitic outbursts succeed each other at short intervals, sometimes without remission, and which are no less dangerous
from their number than from the contagious quality of
their manifestations. In the second place, this time, more
or less lengthened, passed without accidents, permits us to

judge of the degree of amelioration of the diathesis. It is an evidence of the non-activity, of the actual sedation of the diathesis. Without doubt, it is possible that the absence of accidents during a certain period may be only a truce granted by the disease, which, after a certain time, will reassert all its rights. But it is possible, also, in cases where treatment has been vigorously employed, that it may be the commencement of a definitive peace. Why should it begin, if it be not actually a definitive peace?

At all events, it is undeniable that a prolonged immunity constitutes a good sign—a sign which corresponds to an arrest of the disease, which attests its decrease, its decline, at least provisionally. And if this favorable condition be strengthened by the additional guarantee of a sufficient medication, there is reason to hope that the diathesis has definitively imposed silence upon all further manifestations. Security is then acquired for marriage.

So true is all this, that a prudent physician will never permit a syphilitic subject, who has recently had syphilitic manifestations, to marry. For my part, I never allow one of my patients to marry soon after any specific accident whatever, no matter how trifling it may have been. And this for two reasons: first, because the occurrence of any syphilitic accident proves not only that the diathesis continues to exist, but that it exists in full activity; and, in the second place, because, in such cases, it is impossible to predict what may follow. Are not other morbid determinations about to arise in the near future, or is a calm about to ensue? Time alone can decide this question. *To wait* is, then, the rigorous rule in such a situation. On the contrary, if a patient comes to me saying, "It now is two years, four years, six years, ten years since I have experienced anything," this long silence of the disease puts me quite at ease. I perceive that I have to do with a diathesis in process of sedation, with a diathesis which has passed its active stage, which will not reproduce those secondary phases so dangerous as regards contagion, heredity, etc. Consequently, my apprehensions, so far as relates to marriage, are by so much diminished.

I will add that this period of immunity will satisfy me much better still, if it has coincided with a prolonged suspension of specific treatment, for then it assumes a much more significant import; it testifies that the disease, even when left to its own course, and independently of all repressive influence of treatment, has shown no tendency to reproduce its manifestations. And this is not without an important interest for us. We know, in fact, that there are certain cases of syphilis, which, at the same time tractable and rebellious to the action of our remedies, seem to be cured when they are treated, only to burst forth anew soon after the suppression of treatment.* We must distrust this class of cases, and not forget that an immunity prolonged, *without any therapeutic intervention,*

---

* There are, without doubt, certain cases of syphilis in which one can not, so to speak, relax the treatment beyond a certain time without the occurrence of a new outbreak. So long as they are subjected to specific treatment they remain inactive. But let this treatment be suspended, and almost immediately—or, at least, after a certain time—they develop new symptoms. It will be necessary to always continue treating them, in order to keep them quiet. Now, from our present point of view, what guarantee does an immunity due to a *permanent* therapeutic influence afford in such cases? Should we allow a patient affected with a syphilis of this kind to marry, basing our decision upon the twofold guarantee of a long immunity and a protracted treatment? But it may happen that, so soon as the treatment is suspended, new accidents may be produced, with all their consequent dangers. Then, indeed, the marriage should not be tolerated, except in the special condition that a long period shall have passed without accidents, and that, too, *without any therapeutic intervention.*

The forms of syphilis to which I have just alluded are not absolutely rare. It is necessary to distrust them when met with in practice, as long as they preserve their active characteristics and their tendency to continued recurrences.

alone constitutes a veritably substantial guarantee with relation to marriage.

This point stated as a principle—viz., the necessity of a certain period of complete immunity before marriage —I anticipate that here again, as before, you are about to demand the figures of me, to insist upon a precise length of time, and say, "What, then, in your opinion, ought to be the exact duration of this period of immunity, of this sort of probational stage which you require of your patient?"

Now, here again, as before, I can only partly satisfy you. I must confine myself to answering only what I am justified in answering, viz., that—

1. It would be impossible to lay down a fixed, precise term. One must, from the very nature of such questions, depend upon approximative averages within sufficiently wide limits.

2. The longer this period of immunity, the more reassuring will it be in every respect, especially as affecting the question of marriage.

3. Finally, in order to fix a minimum, I conclude, from my personal observation, that it would be imprudent to reduce this period of complete immunity to less than from *eighteen months to two years.* Eighteen months to two years passed without any accident, without any awakening of the diathesis, seems to me a strictly necessary minimum which should be rigorously required of every syphilitic subject before allowing him to marry.

Moreover, let us note well that the duration of this stage of morbid immunity naturally rests subject to variable conditions. It should be longer or shorter with different patients, and the physician's duty will be to proportion it to the exigencies of each particular case. It is evident, for example, that we shall be justified in extending it longer if the most recent accidents presented by the patient have been of a serious nature ; or, again, if the diathesis, in general, has assumed a threatening character. And, conversely, also, we can depart from such a severity in cases where the conditions are directly opposite.

But I perceive that we are about to encroach upon the fourth point of my programme with considerations of this kind. Let us reserve them for the next section.

Fourth condition : NON-MENACING CHARACTER OF THE DIATHESIS.

Evidently there is syphilis and syphilis, as has been so often said and repeated, especially in our day.

There is, unquestionably, a *benign syphilis* and a *grave syphilis.*

There are light, benign cases of syphilis, which, however little they may be subjected to treatment, limit their manifestations to a small number of external, superficial, and unimportant accidents ; while there are, also, grave cases of syphilis, which, although systematically and energetically treated, none the less determine serious and important manifestations—doubly important both on account of the number and of the character of their accidents.

Now, the *quality* (permit me the word), the quality, I say, of the syphilis with which a patient has been affected is far from being without interest in deciding the special question which now engages our attention. Quite the contrary ; it has a superior significance in this matter— a significance very essential to appreciate, and to which there is reason to attach great importance in the solution of our problem.

And, in fact, given the case of a patient who comes to

ask our advice upon the propriety of marriage, if the syphilis with which he has been affected has been only average
or light, if it has been limited to a small number of crops,
if the accidents composing these crops have been superficial and benign, if the diathesis has shown itself tractable
to treatment, and readily and rapidly amends under the
influence of therapeutic agents, we have here, unquestionably, so many excellent conditions which ought to influence our judgment favorably. This *ensemble* of benignity
is well adapted, assuredly, to inspire confidence, and induce. the physician to depart, almost in spite of himself,
from the severity usually necessary in so serious a situation. The past seems here a guarantee of the future, and
compels, if I may thus express it, his acquiescence in the
marriage.

Indeed, this is only justice, for, according to experience, the prognostications of good augury deduced from
antecedents so favorable are almost always confirmed by
subsequent developments, especially when the patient
counts among his assets a long and careful treatment.

Nevertheless, one should not go too far in this direction. Certainly, an original benignity of a syphilis constitutes a favorable condition for marriage, but it constitutes no more than this; it does not, of itself alone,
supply the place of the other requirements of the complex
programme which we are studying. To trust to this
alone, in permitting marriage, would be a grave imprudence, which would lead to the most unhappy consequences; and I regret to say that, in practice, this
imprudence is too frequently committed, the proofs of
which I have in my possession. I shall not hesitate to
insist upon this point, and I repeat that, however benign
may have been a syphilis in its earlier stages, one is not

authorized by this fact alone, by this sole condition, without more ample investigation, without requirements of another kind, to permit marriage. In spite of this benignity, which I have just recognized, and the importance of which, from our present point of view, I again affirm, it is, in my opinion, no less necessary that the patient fully satisfy other general conditions to which every syphilitic candidate for marriage is subject. This is indispensable, and here is the reason:

On the one hand, experience teaches us that cases of syphilis, originally benign, may reveal themselves, more or less tardily, by severe manifestations, if they have not been subjected to an active and prolonged treatment. And it has been often seen that a syphilis of this kind negligently treated, on account of its apparent benignity, becomes later on singularly dangerous in marriage from the double point of view of *contagion* and *heredity.*

On the other hand, and this is no less important, these same cases of syphilis, inoffensive in appearance at first, nevertheless present a dark outlook for the future, and this dark outlook (permit me the expression) relates to the *personal risks of the husband.* I will explain my meaning.

It is now thoroughly proved—(and you will do me the justice, I trust, to admit that I have contributed in part to this important demonstration)—it is to-day well proved, I say, that the initial benignity of a syphilis by no means constitutes an absolute guarantee for the future. Such a syphilis, although beginning well, is no less liable on that account to end badly.* It is thus very frequently that

---

* This is a point upon which I do not cease to insist in all my courses of lectures. I have studied and developed it at length in my *Leçons sur la syphilis chez la femme*, and in this work will be found a long series of observations relative to

one observes patients who, while presenting at the début of the diathesis only slight, almost insignificant secondary accidents, are affected after a long interval—ten, fifteen, or twenty years later, with the most severe tertiary manifestations. Cerebral syphilis, for example, as I have shown in a recent publication,* seems to affect from preference subjects with specific antecedents of an unusual benignity. It is likewise the case, as I hope to show you soon, with syphilis of the spinal cord. And, besides, gentlemen, I am now preaching to those who ought to be converts, for I am speaking to the pupils of the Saint Louis, of the hospital which is the chosen refuge of old syphilitics. Now, do we not encounter here, almost daily, patients affected with tertiary lesions of every character, and with tertiary lesions unfavorable in prognosis local or general, yet whose syphilis commenced the most light, the most benign, the most favorable in appearance? So true is this, that among numbers of them we have great difficulty even in determining the original date of their disease—a date almost forgotten, not only by reason of its chronological distance, but by reason of the trifling importance of the accidents which signalized it.

Those curious cases of syphilis which, after having been characterized by a singular initial benignity, have finally ended in the gravest accidents, have been discussed at length. For my part, I see nothing extraordinary in them ; and I consider them simply cases of syphilis which,

patients who, presenting at the début of the diathesis only the slightest and most benign accidents, have afterward experienced the severest forms of the tertiary stage, p. 1012.

* *La syphilis du cerveau,* clinical lectures recorded by E. Brissaud, Paris, 1879. M. Broadbent professes a similar opinion. "From the cases which I have seen," he writes, "I have formed this opinion, viz., that the subjects most exposed to accidents of the nervous system are those in whom the secondary symptoms have been transitory or light."

having been insufficiently treated, on account of their apparent benignity, have afterward resulted in those accidents which every syphilis not treated may determine.

At all events, from the foregoing facts we may deduce this definite and important conclusion, viz. : *that the initial benignity of a syphilis does not constitute a pledge of security for marriage, if it be not joined with additional guarantees—notably that of a sufficient treatment.* After these remarks upon benign syphilis, we come now to the study of syphilis of a directly opposite character.

There is, I have said, a *bad syphilis* for marriage. This bad syphilis (I retain the word and explain it) comprises all those cases which, from various causes, are more liable than others to become dangerous in marriage.

What are they? They are numerous and of various orders. I can not enumerate them all, for one can not foresee them all, and the particular cases are multiple and infinitely variable. But I will cite here the principal ones, those which it is essentially important to recognize, in order to distrust them, in order to guard against them in the solution of the problem which we are now studying :

1. In the first place, bad for marriage are certain cases of syphilis which, without being severe, present an unusual tendency to continual reproduction—to a ready, repeated, and almost incessant multiplication of various accidents of the secondary stage, notably erosions of the mucous surfaces.

It is thus that certain subjects remain exposed during many consecutive years, sometimes even in spite of the best regular treatment, to erosive lesions, localized especially in the mouth, affecting more rarely the genital mucous surface. These lesions are always superficial, limited, benign ; they are cured with the utmost facility under the influence of cauterization, aided by certain local attention, but they are cured only to be reproduced, to be incessantly renewed. Intrinsically, they are of no importance ; but, for all that, they are none the less dangerous as regards contagion.

Such is the case, for example, of a patient that I have been treating for a long time.

This young man has been affected for five years with a syphilis which one would be justified in classing as benign, since, after the initial chancre, it has never shown itself except by a roseola, a palmar syphilide of light intensity, and buccal mucous patches. He has been under treatment almost from the début of the disease, and quite regularly. Several times he has been subjected by me to an active mercurialization (fifteen to twenty centigrammes of the proto-iodide daily). Now, despite this treatment, despite all my efforts, this patient (who is a smoker, a circumstance essential to note) has not for five years past ceased to be affected with lingual syphilides, with repetitions almost unremitting. I cure him of one crop ; a month or two later a new crop invades the tongue. Then, new treatment, new cure ; then, rapid return ; and so on in succession. In brief, I always cure him, and "always have to begin again," according to his own expression. After long resistance, upon my earnest entreaty, he entirely renounced the use of tobacco. The crops then became less frequent, but did not entirely cease. And, recently, I have again seen him, with the syphilides covering almost the whole dorsal surface of the tongue.

Now, what would have happened if I, trusting to the comparative benignity of this syphilis and the activity of its regular treatment, had allowed this patient to marry in the interval between two crops of these accidents ? I have not to predict, theoretically, what would have happened, for I have had a practical demonstration. This young man last year took for a mistress a young woman until then perfectly healthy and free from every venereal accident. Some weeks later he brought her to me affected with an indurated labial chancre, a chancre evidently derived by contagion from the lingual syphilides of my patient.

2. Equally bad for marriage are the numerous varieties of syphilis which, in various respects, one may characterize as grave : grave it may be on account of the multiplicity and intensity of their accidents (precocious, malignant syphilis, for example) ; it may be on account of the nature of their manifestations (ulcerations, deep, extensive, phagedenic, menacing, etc.) ; it may be on account of their precocious tendency to the visceral form, or, more generally, to morbid determinations which ordinarily occur only in an advanced age of the diathesis ; it may be on account of the reaction which they exert upon the constitution, the nutrition, the health (syphilis of an asthenic, depressive, innutritive form, etc.) ; it may be on account of their character of obstinacy to treatment ; it may be, finally, on account of such or such other peculiarities, infinitely variable, but all presenting this common quality, all unequivocally testifying to an unusual intensity, to a real malignity, even, of the diathesis.

3. Bad again, more particularly from the same point of view, are those cases of syphilis which elect for their morbid determinations some important organ, such as the eye, the brain, the spinal cord, etc.

The cases of syphilis with ocular localizations, for example, are very often remarkable for their obstinacy, for their tenacity, for their relapses after cure, and altogether by the serious functional troubles which they leave in their train. Numbers of times have I seen them result in complete blindness, despite the most energetic treatment and the efforts of the most skillful ophthalmologists.

What is to be said of cerebral syphilis? Every localization of the diathesis toward the brain presages the most serious results both for the present and for the future. To be sure, a cerebral syphilis, even though very grave, can be cured; I have cited examples of this. But, in the first place, how can it be cured? At the price of a most active treatment—of a treatment which must be prolonged, which must be resumed after the cure many and many times; and more, at the price of a special hygiene, upon which I have elsewhere insisted at length, and which requires an observance almost indefinite in duration.* Then, the cure obtained, there still remains the chapter of recurrences, and recurrences in this form of the disease are the most common, and at the same time the most serious. Many a patient, who resists the first assault of the diathesis upon the brain, succumbs to a second or to a third. These recurrences are so habitual that they almost constitute *the rule* †—hence, in all other respects, as well as in relation to the question of marriage, the particularly serious prognosis of every cerebral manifestation dependent upon syphilis.

So, when consulted upon the propriety of marriage by a subject who gives a history of such or such symptoms of cerebral syphilis in his specific antecedents, the physician ought more than ever to be armed with prudence and firmness. In my opinion, from my own observation, every specific manifestation toward the brain constitutes an almost *express* interdiction to marriage by reason of the future

---

* V. *Syphilis du cerveau*, p. 596.   † Vide same work, p. 528.

consequences to which it leaves the patient exposed. For my part, I would most energetically dissuade from all designs of marriage any man who, even though cured of his syphilis, had disclosed to me undoubted accidents of specific encephalopathy in his past history, such as an epileptiform attack, an apoplectiform stroke, hemiplegia, mental affections, etc. Such accidents are, in my opinion, absolutely *incompatible* with marriage. I will not even discuss the supposition of a possible marriage under these conditions.

But if, nevertheless, the diathesis, although positively affecting the brain, has limited itself to the most superficial expressions, to the lightest, most benign functional troubles, then only should I consider myself justified in departing from an absolute interdiction. But, even then, I should not grant my consent until after a mature analysis of the clinical features, and the fulfillment of certain express conditions, such as the following: if the patient is absolutely cured of every cerebral trouble; if he has been a long time cured—that is to say, several years at the minimum; if, since then, no new accident, however slight, has occurred; if the most energetic treatment has been followed since the cure; if a long period of immunity after the suspension of the treatment appears to prove a complete cure, etc. And still—I confess it—even in spite of these guarantees, it would be not without a secret apprehension—it would be only reluctantly, and with a veritable regret—that I should allow, upon my own responsibility, a patient who had been affected with such accidents to engage himself in marriage.

In this matter, gentlemen, you will perhaps say that I am very strict. But I will reply to you, once for all, that we must be strict now or never when it concerns (1) an optional act such as marriage, to which no one is compelled except by his own free will and his individual convenience ; (2) of multiple and important interests involving the future of an entire family. At all events, I do not thus speak to you without being authorized to do so by lamentable and painful examples. Such are the two following, among many others, which I desire to make known to you :

A young man, syphilitic for nine years, having presented only slight specific accidents, and never having been treated, except in a very insufficient manner, suddenly experiences symptoms of a cerebral character. One day, in hunting, he perceives that he can no longer carry his gun in his left hand. His left arm, without being completely paralyzed, has become suddenly inert, benumbed, "half dead." An energetic treatment (mercurial frictions and iodide) is resorted to immediately, and promptly corrects these accidents. The year following, there is a return of the same symptoms. After many repetitions of the same kind, he experiences an embarrassment in the use of his tongue, with sputtering and stammering, and a difficulty in finding and articulating words. A new treatment of the same kind, and all disappears. The patient then writes to me to consult me upon the subject of a marriage which is proposed to him. I strongly advise him to let such projects drop. Notwithstanding, he goes his own way and marries. Now, ten days after his marriage he is suddenly seized with cerebral accidents of the utmost gravity : apoplectiform stroke, hemiplegia, complete amnesia, intellectual troubles, etc. In spite of treatment, all these phenomena persist, and become aggravated. Progressive intellectual depression, general enfeeblement, death from dementia six months later.

Second case, almost identical with the preceding. A young man, syphilitic since 1863, is taken in 1870 with a violent attack of cephalalgia, with incomplete paralysis of the third pair (external strabismus, mydriasis, diplopia). I treat him, and I have the happiness to cure him rapidly. At this time he leaves Paris, and I entirely lose sight of him. In the country he marries, in opposition to the advice of one of my former pupils, consulted upon this subject. Some years later (in 1875) I am again summoned to see him, and I find him in the most lamentable situation : left hemiplegia, amnesia, psychic troubles, hebetude, etc. A vigorous treatment is then instituted, and pursued for a long time. The patient's life is saved, but he remains with his left arm weakened and his intellect enfeebled. No longer capable of managing his affairs, he has liquidated his business, not without great material losses ; so that to-day, with a wife and two children in his charge, he vegetates in a situation the most deplorable, almost allied to misery.

What I have just said of syphilitic affections of the eye and the brain as so many contraindications to marriage, I could repeat literally respecting the specific lesions of the medulla, which also are specially remarkable for their obstinacy, their recrudescences, their recurrences, and which also frequently terminate in the most serious infirmities. To convince you of this, I would recall the case of the unfortunate patient now lying in bed No. 27 of Saint-Louis ward. Affected at three different times with paraplegic symptoms, which depend, from all the evidence, upon an old syphilis, he recovered three times, thanks to an energetic treatment, prescribed in turn by M. Vidal, by M. A. Guérin, and by myself. A fourth time within the past year the same accidents are again renewed, but with an

added intensity. So that, despite a most active medication, notwithstanding all that I can do for him, the patient finds himself to-day in a situation almost absolutely
desperate, even under the menace of a fatal termination
that can not be long delayed.

After these different examples which I have just cited,
it would be useless, I think, to continue this enumeration.
The foregoing facts ought to suffice amply for the demonstration of the proposition which I wish to establish, viz.,
that a certain kind of syphilis, and even a certain class of
syphilitic symptoms, are of a nature to make the physician
very circumspect and very strict in the verdict which he
is called upon to render as to a syphilitic's fitness for
marriage.

One of the essential elements of such a verdict lies, as
we have just demonstrated, in the appreciation of the *in-
trinsic prognosis of each particular case*—in the exact
determination—at least as exact, as precise as possible—
of the *quality* of the syphilis which affects the patient,
who comes to seek our advice and submit his destinies to
our direction.

It is, then, the business of the man of art, in such circumstances, to inform himself as thoroughly as possible
upon his patient's antecedents and upon the nature of the
accidents which he has presented. It is his business to
prepare from a careful and minute inventory what I will
call the *pathological balance-sheet* of his patient, to judge
of the quality of the diathesis under observation ; then,
this analysis made, to decide finally whether, from a medical point of view, there is reason to consider said diathesis
dangerous for marriage or not.

In this matter there are no general rules to lay down ;
for here all depends upon the individual case and the par-

ticular circumstances which surround it, and all remains subject to the knowledge, to the tact, to the experience of the physician.

Here is the true clinical aspect of the problem, and I need not speak of the considerable importance which attaches to it.

CHAPTER X.

Fifth Condition: SUFFICIENT SPECIFIC TREATMENT.—
A treatment sufficiently prolonged, a treatment sufficiently
protective, such is the fifth and last condition of our programme. And this, certainly, is the great condition *par
excellence.* For, after all, in the problem which we are
studying all converges, all reverts to this question : a
syphilitic patient aspiring to marriage, is he or is he not
sufficiently well *cured* of his diathesis to be no longer dangerous in marriage? From this point of view the question of the admissibility to marriage of a syphilitic subject
is almost equivalent to the cure or the non-cure of his disease. We shall not, in consequence, have to spend much
time in demonstrating that which no longer requires to be
demonstrated—that which is to-day accepted by every one,
viz., that the treatment termed specific does in a general
manner diminish and avert the dangers of syphilis.

Whence results this quite natural corollary concerning
our subject : specific treatment diminishes or averts the
dangers of syphilis in relation to marriage. It confers
upon a patient formerly affected with syphilis the most
valuable and the most substantial guarantee of his fitness
for marriage.

This proposition will find its proofs in the several considerations following :

I. In the first place, according to all evidence, specific treatment constitutes the best safeguard, the surest guarantee against the *personal* risks which affect the husband in the relation of marriage.

To be convinced of this, it suffices to compare, in an advanced stage of their evolution, cases of syphilis treated and cases of syphilis not treated.* Exception made of certain cases which baffle all our therapeutic efforts, one may say that a syphilis treated (I mean treated with method, energy, and perseverance) will not have a tertiary period. Beyond a variable number of initial crops, it produces nothing more ; it becomes and remains quiet, and the patient, henceforth exempt from accidents, seems to have been restored to the ordinary conditions of health. Quite on the contrary, syphilis treated insufficiently, or not at all, terminates certainly, constantly, in grave tertiary lesions at a period more or less advanced.

The tertiary period is the time of maturity, when the indifferent, negligent syphilitic pays his accumulated debt to the disease, and "the pox," as M. Ricord has said, "is a pitiless creditor, and grants grace to no one." What examples of this kind have you not here under your eyes ! What of tertiary lesions here in our wards ! And almost all consecutive to cases of syphilis carelessly treated in their early stage, or even (which is not rare) absolutely undisturbed by any treatment.

But we leave the consideration of this point, which, I repeat, is now admitted by every one, save very rare and almost unaccountable exceptions.

* See upon this subject, in my *Leçons sur la syphilis chez la femme,* a long chapter devoted to a parallel drawn between the cases of syphilis treated and those not treated. I think I have there demonstrated—as so many others have done—the inestimable benefits of a systematic treatment, no less than the *disastrous* consequences of a system of expectation when applied to the pox (p. 1052, and following).

II. It is no less evident that specific treatment diminishes and suppresses the chances of contagion in marriage. In effect, a large majority, if not all, patients submitted to specific treatment very quickly acquire an immunity. Look at what takes place in every-day practice. A patient comes to us in the full secondary stage, and we subject him to the usual treatment. What is the result, at least nineteen times out of twenty ? In the first place, the patient remains subject during the course of the first few months, sometimes even during the first year, to secondary crops, more or less numerous, more or less intense, according to the quality of the diathesis, but, generally, they are lessened and rendered milder by the treatment. Then, afterward, from about the second year, the crops commence decreasing ; they are limited to a few isolated and benign manifestations—for example, to certain buccal erosions. Then, again, later on, the calm becomes more pronounced ; it becomes complete with the third, or, later, with the fourth year. After that time it is all over with the secondary period ; and with it are finished those contagious accidents which it comprises, and which constitute the principal danger relative to marriage. Such is the rule.

That there are exceptions to this rule I know only too well, and I have given examples of them previously (vide page 104). But these exceptions are always rare ; and, moreover, they belong to that class of cases which I have pointed out to you as constituting the contraindications to marriage.

III. In the same way, finally, specific treatment diminishes and suppresses the *hereditary* risks of syphilis.

In the first place, this is superabundantly demonstrated in the case of paternal hereditary syphilis. Recall, as examples, those very convincing cases of which I have previously spoken to you, and which may thus be recapitulated : A healthy woman aborts many times in succession without cause and without explanation. We become uneasy ; we search for the wherefore of these successive abortions, and we find no other possible explanation than the syphilis of the husband. Empirically, the husband is then submitted to a severe specific treatment. New pregnancies occur, and these terminate altogether favorably— that is to say, they produce at full term healthy children. What more convincing ?

Now, this favorable effect of treatment is no less evident upon both maternal heredity and the mixed heredity of the parents. And here you will pardon me a short digression, which, though separating us a moment from our present line of inquiry, will shortly find its application.

Specific treatment, I say, corrects with equal effectiveness the influence of *maternal heredity.* In proof of this, innumerable cases are seen where syphilitic women begin by having several abortions, or by bringing forth syphilitic children ; then, after having been submitted to a specific medication, they bring forth at term living and healthy children. The cases of this kind are so common and so numerous, that I think in truth it would be useless to stop here for particular citations. *

Sometimes, again, the influence of treatment upon maternal heredity reveals itself in a much more striking manner, on account of the peculiarities of certain particular cases. Such is, for example, the following case : A

* Example: A woman contracts syphilis, and then has eight miscarriages, without being able to carry an infant to term. She submits herself to a prolonged mercurial treatment, again becomes *enceinte*, and is delivered at term of a well-developed child, which is now five years old, and has never exhibited the slightest trace of syphilis (Notta, memoir cited).

young woman receives syphilis from her first husband, and is treated in a very ephemeral fashion. Left a widow, she is married again to a healthy man, and by this husband conceives several children, that either die *in utero*, or are born syphilitic. She is then treated, and after treatment she brings forth only healthy children.

In the third place, the influence of treatment upon *mixed heredity* is still more often confirmed in practice.

Not unfrequently the following order of events is encountered : Two syphilitic parents begin by engendering a succession of children, all of which either die before birth or are born syphilitic. The parents are then treated. Consecutively to this treatment they procreate other children, which are born at term, living and healthy.

One may often trace in a series of consecutive pregnancies the progressive influence of treatment — when each pregnancy is marked by a step toward a cure. I have collected many facts of this kind, among others the following :

A young man marries, notwithstanding a still recent and very negligently treated syphilis. His wife, almost immediately infected, aborts some months later. The two parents then commence to be treated in earnest. There succeed four pregnancies quite close to one another, which terminate as follows :

1. Accouchement before term ; child still-born.

2. Accouchement at term ; child syphilitic, dying some days later.

3. Accouchement at term ; child syphilitic, but surviving.

4. Accouchement at term ; child healthy.

But this is not all ; and here should be noted a fact of twofold significance, which has not been as yet, it seems to me, sufficiently remarked :

1. It is not necessary, in order that syphilitic parents procreate healthy children, that the diathesis be annihilated in these parents. In other terms, and to speak more clearly, it may happen that the offspring of syphilitic parents may be born healthy, *although their parents are still under the impress of the diathesis,* which is proved by the appearance of specific accidents upon them subsequent to the birth of their children. This is incontestable, so far as relates to the father, as we have previously shown.*

It may be likewise demonstrated as regards the mother. As an example, see one of our present patients, in bed No. 31, in St. Thomas's ward. This woman entered the hospital affected with a sclero-gummatous glossitis, the origin of which dated back three or four months. Now, her last child, aged fourteen months, has never presented the least suspicious accident; it is a very fine child, absolutely healthy, as you have been able to, and may again, convince yourself by sight.

Likewise, a young woman among my patients, who became syphilitic from contact with a syphilitic husband, has had two children absolutely healthy, although after each one of her confinements she has been affected with a very intense eruption of squamous syphilides.†

---

* Vide page 36, and Illustrative Cases, Note I.

† Here is a summary of this curious case: X., twenty-two years old, vigorous constitution. At twenty years married a man affected with a recent syphilis. First pregnancy in 1868. Secondary accidents toward the fifth month (erythemato-papular syphilide, buccal and vulvar syphilides, alopecia, cervical adenopathies). Energetic treatment with mercury; accouchement at term; child healthy; continues so ever since. Two months before accouchement, psoriasiform annular syphilide, occupying the lateral surface of the left foot. Ecthyma of the left leg; mercurial and iodide of potassium treatment.

Second pregnancy in 1872; accouchement at term; child healthy; continues well to this day. Three months before accouchement papulo-squamous, psoriasi-

9

2. In order that a child be born healthy of syphilitic parents, it may suffice that these parents are being subjected to mercurial influence at the time of its procreation. However singular, however paradoxical, and, above all, however inexplicable, such a fact may at first glance appear, it would seem to be evident from a certain number of well-authenticated cases. Such is, for instance, a case mentioned by Turhmann (de Schoenfeld), and which is briefly as follows : A syphilitic woman first of all has seven pregnancies, during which she is not treated. Seven times she gives birth to syphilitic children, which soon die. Becoming pregnant an eighth and a ninth time, she is treated during the course of these two pregnancies, and each time she is delivered of a sound, healthy infant. A tenth pregnancy ensues ; this time the patient is not treated. She is delivered of a *syphilitic* child, which dies at six months. Finally, an eleventh pregnancy, in the course of which treatment is resorted to, results in a *healthy* infant.*

Had this case been fabricated out and out, imagined theoretically to meet a necessity, it could not, in truth, have been more convincing.

For my part, I have already among my notes certain observations of the same kind relative to syphilitic parents who have alternately engendered healthy children, at a time when they were subjected preliminarily to a specific treatment, and syphilitic children during a period when they were not being treated.†

---

form, circinate syphilide, constituting a ring of large diameter upon the dorsal aspect of left foot. Resumption of treatment; cure of the accidents.

* Vide *Gazette Médicale*, June 24, 1843.

† The same fact has been likewise remarked by M. Kassowitz (*Die Vererbung der Syphilis*, Vienna, 1876). Among a number of observations testifying in the same way, I will cite the following, which is accompanied with all possible guarantees

It would seem, then, from this, that even the temporary influence of treatment may suffice to avert for the time being the effects of syphilitic heredity. Such, at least, would be the conclusion to be drawn from the preceding cases. But I am not willing to give you this last conclusion as an absolutely demonstrated fact. More prudently, I restrict myself to presenting it to you as a subject for further study—very interesting assuredly, already rendered probable by certain observations apparently well authenticated, but still lacking in a sufficient number of guarantees to assure it the right to be cited as a scientific fact.

All the considerations which precede concur in many ways in establishing the modifying, corrective, depurative influence which treatment exerts upon the diathesis. And the natural conclusion which results from all this, as far as we are specially concerned, is that—

The essential, capital condition to be fulfilled by every syphilitic subject aspiring to marriage consists

of authenticity : A young workman marries in the third year of a syphilis quite vigorously treated. His wife becomes *enceinte* after some months, and toward the end of the first half of pregnancy begins to present symptoms of secondary syphilis (roseola, neuralgias, mucous syphilides, alopecia, etc.). She is submitted to a very active treatment, and accouches at term. The infant is born syphilitic, and dies of consumption when one month old. The young wife continues to be treated and becomes *enceinte* five months later. She is delivered of a fine child, which, very carefully observed, remains exempt from every suspicious phenomenon, and is to-day in perfect health. Reassured as to the condition of this woman by the fact of her giving birth to a healthy child, her physician does not treat her any more. She becomes *enceinte* a year later and aborts. New pregnancy some months later; another abortion. Two years later a fifth pregnancy ensues ; birth at term, of twin children, both syphilitic.

Even supposing that the influence of treatment gives occasion for dispute in this case, the fact no less remains established, by this and similar observations, that a *syphilitic woman may alternately engender syphilitic children and healthy children.* This is an actual fact, which forces itself on the conviction, however strange and paradoxical it may appear. It remains for us to find its explanation ; but it is only upon the explanation, and not upon the fact itself, that there can be differences of opinion.

in a thorough specific treatment, in a treatment sufficient to confer a complete immunity from the multiple and diverse dangers which syphilis carries with it in marriage.

In order that a syphilitic patient may have the moral right to become husband, father, and support of a family, it is necessary, it is indispensable, that he shall, by virtue of a treatment sufficiently protective, have ceased to be dangerous to his wife, his children, and himself.

But what is this treatment, *"sufficient, sufficiently protective,"* to which we are constantly referring as our best safeguard?

This, gentlemen, I have gone over with you at length in a former series of lectures, tracing for you in detail the rules for the treatment of syphilis, such, at least, as I comprehend them; such as they have been taught to me both by my masters and by my personal experience.* At present, then, I have only to refer you to these lectures, a portion of which has already been given for publication, and which you may consult at leisure. I will limit myself to-day to only reminding you in quite a brief manner that a treatment worthy to be characterized as "sufficient" is this:

1. A treatment which is based upon the administration of the two great remedies which, with just reason, are commonly called the "specifics for the pox," viz., *mercury* and *iodide of potassium.*

2. A treatment which is based upon the administration of these two remedies in *doses veritably active and curative,* very different from the insufficient, timid, indiffer-

---

* *Du traitement de la syphilis* (lectures delivered at the Saint-Louis Hospital). In press.

ent, almost inert, doses in which, according to traditional routine, they are most often prescribed.

3. A treatment which is prescribed and regulated according to a certain method, which has for its aim and result to conserve for these remedies, notwithstanding their prolonged administration, their primitive intensity of action (called the *system of intermittent or successive treatments*).*

4. A treatment which, in these conditions, is rigorously pursued during *several consecutive years*—at the minimum, *three* or *four* years.

I especially insist upon the importance of this last point, and I say, *a chronic disease*, in effect, *requires chronic treatment*. Such is the absolute general law. Long, very long, should be the medication, if we are not content alone with a present effect ; if we desire to obtain a complete, thorough, permanent cure.

According to experience, it is false, absolutely false, that one has "finished with the pox" after a treatment of some months, of one year, of two years even (the extreme limit which is scarcely surpassed ordinarily). Treatment of this kind furnishes nothing more than a *provisional* immunity, a temporary silence of the disease, leaving the diathesis in full force, with all its future dangers, with the

---

* Vide *Leçons sur la syphilis étudiée plus particulièrement chez la femme*. Paris, 1873, p. 1087, *et suiv*. A fact, of which experience has absolutely convinced me, is that mercury and iodide of potassium, when administered for a long time, and without intermission, singularly lose their efficacy. The continued use of these two remedies, as well as of many others besides, creates a tolerance which enfeebles, lessens, and ends by annulling their therapeutical effects. On this account I was forced to devise for my patients a method of treatment which should conserve the full intensity of therapeutic action of both the mercury and the iodide during the entire duration of treatment. This method I have exposed at length in my teachings under the name of *the method of successive or intermittent treatments*. I think I am authorized in saying that it has been of real service in practice, and I commend it to the attention of my *confrères*.

certain imminence of tertiary accidents at a subsequent period. Treatment of this kind, I may say, is to-day condemned by its numerous deplorable results. It is, indeed, high time to renounce, once for all, these *curtailed* medications, to approach the pox from the same therapeutic point of view as we do other constitutional diseases, such as scrofula, gout, malaria, etc., which are, by common consent, curable only by a long-protracted treatment, only by a series of successive cures, only by the repeated, almost chronic, intervention of the remedies appropriate to combat them.

For my part, I consider myself authorized to assert, from my own experience, that in *no* case should the duration of an anti-syphilitic treatment fall below three or four years, with whatever form of the disease we may have to do, however benignly the diathesis may have been originally announced. Three or four years methodically devoted to an energetic medication—such, in my opinion, is the necessary *minimum*, I will not say to cure the disease, but to avert its dangerous manifestations both for the present and for the future.

Again, it is prudent that, after this term, the patient submit himself from time to time—every two or three years, for example—to a new iodide treatment, so as to keep the diathesis incessantly in check, if I may thus express it, and to hold the ground gained. Combined with time, the specific treatment of the diathesis constitutes certainly the *best guarantee* in favor of the syphilitic subject who aspires to marriage.

Time, on the one hand, treatment, on the other, are unquestionably the two grand correctives of the pox, the two essential conditions to require of every syphilitic subject before opening to him the gates of marriage.

CHAPTER XI.

I CAN not quit this subject of treatment without adding some remarks relative to a custom very fashionable, and considered by the public as an infallible criterion of the cure of syphilis.

A popular belief, as you know, ascribes to the sulphur mineral waters the singular property of revealing, of "bringing out," the pox in syphilitic subjects not yet cured of their disease.

With this object in view, numbers of patients journey each year to such or such sulphur spring, either at their own suggestion or by the advice of their physicians. There they religiously take the waters during the traditional twenty-one days, awaiting, not without anxiety, the result of this treatment. According to their theory, "if they have anything in the blood the waters will bring it out; while, if they are cured, if they have nothing, nothing will come forth." In the first alternative, the appearance upon the skin of new syphilitic symptoms will be an indication that they must undergo fresh treatment; while, in the second, the absence of external manifestations will constitute a pledge of cure.

Now, this "*judgment of the waters*" has been applied (and it could not fail to be) to the serious question of mar-

riage. You will find this opinion entertained by numbers
of your patients, that, before thinking of taking a wife, it
is the duty of every syphilitic subject to make a pilgrim-
age to some sulphur spring, in order to ascertain what
pertains to his specific state in general, and his matrimonial fitness in particular.

Now—have I need to say it?—this pretended *revealing
action* of the sulphur-waters is far from being what it is
gratuitously supposed to be. It is very far from the truth
that it "*unveils the unknown,*" following the classic expression, and that it furnishes a sort of criterion of the
cure of syphilis.

Without doubt, sulphur-waters may determine specific
eruptions in syphilitic subjects in some cases ; and it could
not well be otherwise, considering the excitant, irritant
action which they exert upon the skin, especially when
used daily in the form of baths, douches, vapor-baths, etc.,
as is the case in the great majority of our bathing stations.
All hydropathic physicians have observed and reported
cases of this kind, and I could myself relate several
examples.

But this action of the waters, we must remark, is not
constant; and here is the proof: In the first place, numbers of syphilitic patients are sent each year to sulphur
stations from divers motives. Even when they are supposed to be by no means cured they are sent there, for
example, in order to be built up, in order "to recruit both
from their disease and from the effects of treatment."
But we observe that they almost all return from these
thermal stations without having experienced the least new
manifestation upon the skin, without having perceived the
least cutaneous awakening of the diathesis. On the other
hand, we have now the experience of these so-called *reveal-*

ing cures, and we know what they are worth. I have in my notes hundreds of observations relating to patients who have taken one, two, three, and up to six courses of sulphur-waters without having had any eruption produced upon them, and who afterward, at quite distant intervals, have suffered from several severe assaults of the disease. Within a few days, for example, I have been called to treat a man still young, presenting undoubted symptoms of cerebral syphilis. Now, this patient, like so many others, had gone to Luchon before his marriage, and on account, even, of his projected marriage. He had been there three seasons, and nothing had been manifested upon him. After such assurance, it was thought that he might be allowed to marry without apprehension. The result has shown the justness of the prognosis!

A similar result has occurred likewise in many and many other cases which I could cite, and of which every physician could enlarge the list.

The revealing action of the sulphur-waters constitutes, then, in no respect a *criterion* which can be relied upon. The test of the waters is a legend to be abandoned like so many others. It is false, absolutely false, that the sulphur-waters "disengage" the pox from the organism in the manner of a reagent, which disengages a body from a chemical combination. And, clinically, we can expect no solid guarantee from a thermal treatment in order to determine the question of the cure of our patients who are contemplating marriage.

For one time that this revealing action takes place, it will fail twenty times, fifty times, perhaps. What security can a process so subject to failure furnish? What would be thought in chemistry, for example, of a reagent which, nineteen times out of twenty, did not disclose the special body which it was intended to reveal?

And let no one accuse me of making a war of prejudice against sulphur-waters. The accusation would fall to the ground, as I am "a believer" in sulphur-waters—I prescribe them frequently, quite frequently, in the course of syphilis, and every year I send numerous patients to our stations in the Alps and the Pyrenees. But I believe in these waters, and I prescribe them, for other indications than that of their pretended revealing property. I believe in their utility as tonic, renovating agents, especially in cases of syphilis of the asthenic form, or of syphilis complicated with lymphatism, scrofula, etc. I believe also that they may render incontestable service in facilitating the tolerance of strong mercurial medication in those cases where there is occasion to require of mercury its full intensity of action. I do not deny that they may sometimes come in as an aid to diagnosis, in determining cutaneous manifestations upon our patients which could not be produced without them. That only which I deny, which I energetically repel as a dangerous error, is their pretended *faculty of arbitration* in questions so grave as these—the cure or the non-cure of the pox—the fitness or the unfitness of a syphilitic subject for marriage.

I have just passed in review before you, gentlemen, the several conditions which, in my opinion, a syphilitic subject should satisfy in order to justify him in aspiring to marriage.

From the foregoing, you have already deduced these natural conclusions:

1. To every patient not fulfilling the conditions—*all* the conditions of this programme—I believe that the physician ought expressly and energetically to interdict marriage.

2. To every patient satisfying fully and completely all these conditions, I believe that the physician may safely *permit* marriage.* This, in effect, is only the necessary deduction from the premises which we have established. It is the application of them, pure and simple.

However, after having traced and defined for you this programme of admissibility to marriage, such as I understand it, I should like to add a few reflections, a few comments which appear to me indispensable.

In the first place, take this programme only for what it is and what it is worth. It is not a programme received, discussed, and accepted by contemporary science. It is purely and simply the condensed result of my personal observation, aided by certain contributions which I have been able to gather here and there from diverse sources.

It is, in the second place, a programme subject to revisions, susceptible of amendments, additions, and corrections, and which I shall be the first to modify as soon as further observations shall point out the changes to be in-

---

* I regard it as superfluous to speak here of the various recommendations with which the physician should accompany his acquiescence in the marriage of a syphilitic subject, and which are quite naturally suggested by the preceding considerations. Of these recommendations, the principal and most indispensable one is that which relates to the minute and careful watch which the future husband should exercise over his own person, so as to allow nothing to pass unperceived which could give rise to an offensive return of the diathesis. It is very essential that we should warn our patients of the possible dangers of every lesion which manifests itself upon them, however minute and however insignificant it may appear to be. It is very essential that they receive from us the explicit injunction to refrain from all relations, from every contact, in case they should be affected with any lesion whatever, either of the genital organs or of the mouth, the throat, etc. How often have I heard such or such a patient, who had had the misfortune to infect his wife, bitterly censure his physician because he had not sufficiently enlightened him upon the dangers of contagion! It is always the same complaint which husbands repeat to us in these unfortunate situations: "I was not forewarned. If I had known what I have since learned to my cost, I should never have communicated the disease to my wife," etc. Let us be warned in our turn, and not risk the possibility of such a reproach.

troduced, and certain points of the question, still obscure and inexplicable for me, shall have been elucidated.*

Besides, gentlemen, do not delude yourselves with the possibility of ever constituting that which I would call (permit me the phrase) *a perfect code of marriage, for the use of syphilitic subjects*—that is to say, of building up a programme which shall respond to all possible contingencies, which, in a manner absolutely certain, shall determine in all cases the fitness or the unfitness for marriage of a patient tainted with syphilis. A categorical solution of a problem of this kind, a solution stamped with mathe-

---

* To speak only of one of these points, which I can not explain, and which is to me a source of perpetual astonishment, viz., that certain cases of syphilis show themselves so dangerous, so pernicious, and certain others so benign, so inoffensive, as regards their consequences in marriage. I have in hand certain observations relative to patients who have married in spite of a severe and insufficiently treated syphilis, and have, nevertheless, remained inoffensive both for their wives and for their children. Such is the history of one of my patients who married in spite of me, when scarcely cured of a guttural phagedena of the most menacing character. Such is the case of another patient who, without consulting me, contracted marriage when scarcely cured of multiple accidents of a malignant syphilis (profound ecthyma, rupia, cephalalgia, hemiplegia, etc.). Now, these two imprudent syphilitics, contrary to all rational previsions, have had healthy children, and their wives have remained uncontaminated. Conversely, we sometimes see patients who, although having been affected with only an average or benign type of syphilis, and treated for a more or less protracted time, after conscientiously waiting several years before contracting marriage, have, nevertheless, procreated syphilitic children, and, in one way or another, infected their wives. As an example, I will cite the case of one of my present patients. In 1864 he contracted syphilis, which manifested itself by slight accidents (mucous patches of the throat, crusts of the hairy scalp, temporary falling out of the hair, etc.). He was treated during two or three years. Having been exempt from every suspicious symptom for about five years, in 1871 he thought that he might marry—not, however, without having taken the advice of one of our eminent *confrères* upon this subject. And then he procreated a syphilitic infant, which communicated the syphilis to its mother *in utero!* What contrast, what disparity between these two orders of cases! Both, assuredly, have their *raisons d'être*, their material, organic explanation. But we are forced to confess that in the present state of our knowledge this explanation absolutely baffles us. There is, assuredly, an unknown quantity which eludes us, at least for the present, and doubtless numerous observations will yet be necessary to disengage it from the multiple elements of so complex a problem.

matical precision, is not, and never will be, possible. To even hope for it would be evidence of a non-professional understanding.

Always, whatever we may do, our verdict will be based upon a simple *calculation of probabilities*—that is to say, upon an appreciation, essentially difficult and delicate, of vague and ill-defined elements, such as, on the one hand, the previsional diagnosis of a diathesis, and, on the other hand, the degree of corrective, preventive influence exerted upon this diathesis by treatment. Consequently, we should not conceal it from ourselves, nor, for that matter, from our patients, that our decision can only have a degree of certainty proportionate to the elements which serve as its basis. That is to say, whatever the attention, whatever the painstaking care we may bring to the examination of a particular case, it is not impossible that the results may baffle our expectations; it is not impossible that we may fall into an error. For, I repeat the expression by design, we rely upon, and we can only rely upon, "a calculation of probabilities" in forming our opinion.

And yet, gentlemen, do not get a wrong impression from this last sentence. Because a mathematical certainty fails us in this matter, it does not follow—far from it—that the physician may not be competent to render to patients and to society frequent and inestimable services in this grave matter of the marriage of syphilitic subjects. Consider, for a moment, how these things present themselves, and appreciate exactly the situation as it occurs in practice.

A patient consults his physician in order to know if he may or may not marry, notwithstanding an anterior syphilis. The physician interrogates and examines this patient, searches out the conditions favorable or unfavorable for marriage—in a word, prepares a balance-sheet of the case, and endeavors to form an opinion. One of two things then happens, viz. : either the physician will have gathered from this examination facts sufficient to satisfy him as to the condition of his patient, and enable him to pronounce an opinion in one way or the other (and, in this case, there are but few risks that his judgment will be discredited by the results) ; or, indeed, he will fail to find the necessary elements upon which to found his opinion, to satisfy his judgment. In this second alternative, he will refrain from announcing a verdict. For, be it understood, the physician is not bound even to have an opinion ; he is not in the position of a judge who, from his high tribunal, is under obligation to decide, to pronounce judgment between two adverse parties. On the contrary, he has the opportunity of excusing himself, of appealing to the future, and of saying to his patient: " Under the present conditions it is impossible to know how to act in your position. You might perhaps still be dangerous in marriage. For the present, then, we will not come to a decision ; *it is necessary for you to wait.*"

Such is the situation, and not otherwise : please remark it, for one is too apt to forget it. And, I again repeat, the physician is not obliged to fall either into an error, or to risk a hazardous judgment, when he has not the elements necessary for the solution of the problem before him. It is by remaining faithful to this rule of conduct, by abiding within these limits, within this boundary, that the physician will respect science, and will most effectively serve the interests of his patient.

I have often seen patients marry against the advice and express prohibition of their physician. And, upon consulting either my recollections or my written notes, I find this : that if some few of these imprudent individuals have not had cause to repent of their rashness, yet a very large number of them (I may even say an enormous majority) have experienced the most lamentable catastrophes, either by introducing the pox into their homes, or in procreating children syphilitic or puny, and almost always doomed to an early death; * or, finally, in paying their personal debt to syphilis to the great detriment of their family.

On the other hand, also, I have seen patients marry after a medical examination, and with the consent of their physician. And here, again, the same proportion was met with in the results, but in a directly opposite sense. Save certain very rare, really exceptional cases, where the anticipations of the physician were negatived by the developments, the almost absolute rule was that these patients were not dangerous subsequently, either to their wives, to their children, or to themselves.

---

* I have just observed a case which well merits, by way of example, to be recorded here: A young man contracts the syphilis. At its outset he only experiences slight accidents—roseola, erosive syphilides of the mouth, crusts of the hairy scalp, cervical adenopathies, etc. He is treated with mercury during several months. Everything disappears, and he considers himself cured. About six months later he marries, *in spite of the explicit prohibition of his physician.* His young wife becomes *enceinte* almost immediately. Toward the fourth month of her pregnancy she begins to present unmistakable signs of secondary syphilis (erythemato-papular syphilide, vulvar and buccal syphilides, iritis, cephalalgia, lumbago, neuralgias, sleeplessness, nervous accidents, etc.). She aborts at the sixth month. The following year two other pregnancies ensue—abortion at fifth month; accouchement almost at term of a syphilitic infant, which succumbs in twenty-four hours. Fifteen months afterward, fourth pregnancy : accouchement at eight months of a still-born infant. Treatment always irregularly followed by the husband and by the wife. Ten years later, the husband is attacked with cerebral accidents, which, by common consent of several other physicians and myself, are referred to a specific encephalopathy. Tardily administered, the anti-diathetic treatment only succeeds in temporarily suppressing the accidents, and the patient is rapidly carried off.

The marriages contracted under these conditions have almost invariably terminated in happy results. This I 'am in a position to vouch for, figures in hand. Is it enough to say, then, that the judgment of an intelligent and prudent physician affords in this matter, even from this "calculation of probabilities," substantial guarantees?

Is it enough to say, finally—and I do not revert without satisfaction to this last point—that the physician consulted upon the question of the marriage of a syphilitic patient renders a most important and valuable service by protecting, in this solemn moment, on the one hand, the interests of his patient, and, on the other hand, behind this patient, the interests of society?

PART II.

AFTER MARRIAGE.

CHAPTER XII.

HUSBAND SYPHILITIC AND WIFE HEALTHY.

WE are about, gentlemen, to continue and complete our study of the relations of syphilis with marriage.

Hitherto we have considered the question only *before* marriage. It now remains for us to study it *after* marriage.

The evil which we desired to prevent is accomplished. A syphilitic man, not cured of his syphilis, has been married. He is a husband.

What dangers may result from this situation? And what rôle have we to play, medically, in order to avert or lessen these dangers? Such is the question which it is now necessary for us to approach.

A practical question, if there ever was one; a question teeming, as you will see but too well in the course of this exposition, with embarrassments and difficulties of many kinds, with situations equivocal, delicate, complex, etc.; a question seldom to be met with in the hospital, but of frequent occurrence in private practice, where it forces itself upon the physician's attention. Consequent-

10

ly, I have thought it would be useful to discuss it before you in detail, in order to spare you an apprenticeship, always more or less painful to undergo, in personal experience.

The evil is consummated, I said. A syphilitic man, although not cured, is married, and is now a husband, with a syphilis in full vigor, in full activity of manifestation. A deplorable situation, by no means uncommon, from various causes. It may be that the patient (as is most commonly the case), believing himself cured, inconsiderately contracts a premature marriage ; it may be that, consciously and voluntarily, he braves the dangers of such a position ; or, it may be that, ignorant of his real disease, he mistakes the nature of the lesions with which he was affected before his marriage. And, if you will, we shall place side by side with facts of this kind two other classes of cases which, though very different from the first, as regards the chronological order of the accidents, are no less liable to terminate in a situation exactly identical, viz. :

1. The cases (very numerous these are) in which a married man contracts syphilis *after* his marriage in an extra-conjugal adventure, or rather misadventure.

2. The cases (infinitely more rare, really exceptional) where syphilis breaks out upon a man quite recently married, from the fact of a contagion to which he was exposed some days before marriage. I will explain by an example borrowed from my notes :

A young man, of high social position, has connection, eleven days before his marriage, with one of his former mistresses, who is affected at the time (as is afterward demonstrated) with mucous patches of the vulva. He marries absolutely healthy in appearance. Eight days after his nuptials he notices a slight redness upon the glans, which soon degenerates into a typical indurated chancre, the beginning of a constitutional syphilis, which he soon transmits to his young wife.*

Whatever be the chronological origin of syphilis, whether the contagion be anterior or posterior to marriage, the situation, I repeat, is absolutely the same in these diverse cases. We always find the same scene recurring with the same personages, viz., on the one hand, a healthy woman; on the other, a syphilitic husband. These conditions existing, what will take place? What will take place, first of all, is that this husband, on the first invasion of a suspicious phenomenon, will rush into the office of a physician, and, burning with excitement and anxiety, will address him in the following language (which I reproduce from reality, you may believe me): "Doctor, save me! I believe I have symptoms of syphilis. Now, *I am married.* Just consider a moment my situation, if I were to give the pox to my wife—if I were to have syphilitic children! Preserve me from this, I beg of you, and explain to me everything I should do in order to guard against such dangers."

Consulted under such circumstances (and you will often be, I assure you), what will you reply?

In my opinion, your professional rôle is all traced out; and, if you will be advised by me, your reply will be the following:

"I see *three orders of dangers* in the situation for which you do me the honor to seek my advice, viz. :

"1. Your own personal dangers, those which may result to yourself from your disease.

---

* Cases of this kind are extremely important to know in practice. I have called the attention of my readers to them (vide " Illustrative Cases," Note IV).

"2. A danger of contagion for your wife.

"3. A danger of heredity for your future children.

"Now, it is not only necessary that we should guard against these three orders of dangers ; it is necessary, too, for us to attend to them all in an equal measure ; for you would be culpable, and I should be culpable with you, if we aimed only at your individual security, without taking precautions for your wife and your children. In consequence, it is a triple consultation that I have to give you.

"And, first, let us consider the most urgent indication. Let us speak of yourself. That will be my first step.

"As to you, sir, you must be treated in the most energetic manner, and everything be done in order to cure you at the earliest possible moment. For it is from you, in brief, that all the consequences proceed which may fall upon others.

"To this end, here is my advice," etc.

But, before going further in this scene, what are you about to advise this patient to do—this syphilitic presenting himself to you under these special conditions?

Certainly, as regards the nature and the character of the remedies, you will prescribe for him what you would prescribe for any one else, for there are not, that I am aware of, any special remedies for the particular use of syphilitic husbands.

But as regards intensity, as regards vigor of medication, it is an altogether different affair, in my estimation. Bear in mind that you are in the presence of a *husband*, of a husband living in contact with a young wife, liable to infect this wife in the thousand intimacies of domestic life, without even speaking of sexual relations, from which, it is to be feared that, notwithstanding all your warnings, your patient will not abstain. I will add to this last statement—and it is well to know it in practice—that in such cases your patient will show himself much less tractable to your prescription of continence, as he has an interest in concealing his condition, and is not willing, as he will tell you, "that his wife should divine his disease," etc. Be less sure of him in this particular than you would be of any other patient in a different situation.*

Now, from these special conditions are practically derived special *indications*, of which you will comprehend in advance both the object and the useful results.

These indications, even speaking of the principal ones only, are briefly as follows:

I. In the first place, *to suppress forthwith the sources of contagion*, and to suppress them by a cauterization at once sufficient and energetically corrective.

If, for instance, the case be one (by far the most common) of secondary accidents of the mouth, of the throat, or of the penis, etc., destroy them immediately by a vigorous cauterization. As for the nitrate of silver, a feeble caustic, there might be risk of its proving insufficient; choose, in preference, the acid nitrate of mercury, a more active caustic, and much more certain as regards effects.

Should this cauterization not succeed in correcting immediately, *in situ*, the contagious nature of these accidents, it will at least certainly provoke their cicatrization in a short time ; and this is what we desire.

---

* I have known syphilitic husbands to *not dare* abstain from intercourse with their wives during a period when the contagion was transmissible, from the fear that a suspension of habitual relations might give a hint as to their disease. In order to avoid suspicion, they risk transmitting the pox to their wives! The fact is scarcely credible, but it is none the less authentic, nevertheless. It is authentic to the extent that two of my patients, in conditions of this kind, have, *in endeavoring to prevent suspicion*, ended by infecting their wives. Facts of this kind ought to be signalized, for one would not imagine them *a priori*, one would not suppose them possible, without having had peremptory proofs from personal observation.

It is superfluous to add that the cauterization should be immediately followed up by the employment of such topical applications as are best adapted to promote the speedy cure of those centers of contagion.

II. *To cut short, by a medication of especial intensity, the contagious accidents of the secondary stage.*

Under ordinary circumstances, when called upon to treat syphilis, the medication which we usually prescribe to patients is at once mild, guarded, cautious, careful, our endeavor being to adapt it to individual tolerance. We go slowly, gently, patiently, for we have time before us; and proceed with moderation, even allowing the diathesis to get the upper hand from time to time in temporary outbursts. But, under the special circumstances now under consideration, the case is different. It is urgent here to avert the imminent dangers of contagion. With this view, the indication is to *make haste* and to *strike hard*, if I may thus speak, in order to silence all menacing manifestations—menacing, be it well understood, not for the patient, but for his wife, whom it is necessary to protect.

So, then, in place of the usual medication, in place of the traditional "five centigrammes" of proto-iodide, I believe that there is every advantage in instituting from the outset an energetic repressive treatment. I believe that we should act in this emergency as we act when in presence of grave specific accidents which it is important to suppress promptly. In a word, I am here in favor of a swift and violent treatment, yet avoiding extremes, and without running the risk, in going too fast, of being forced to go backward—I mean forced to suspend the medication. With this intention, prescribe, then, from the very first, strong mercurial doses. Ten to fifteen centigrammes of the proto-iodide, two or three, and even four, centigrammes of the sublimate, daily, would not constitute an excessive average, at least in general, exception made for individual tolerance, which is always to be taken into account. Often, even, it will not be injudicious to associate the iodide with the mercury, in order to render the medication still more active. Pursue this treatment about two months, then discontinue it for some weeks, in order to avoid the effects of habituation, beginning again afterward in the same way, for the same time, and so on successively.*

Proceeding in this manner, you will often, if not always, succeed in suppressing all or a part of the secondary manifestations ; you will succeed notably—and this is the object you have in view—in diminishing the number and intensity, even perhaps in averting completely, the crops of eruption upon the mucous surfaces, which, under the name of mucous patches, are so formidable as regards contagion, and constitute the common, habitual source of contamination in marriage.

Without doubt, this desperate treatment will not always be to your patient's taste. Without doubt, it will cause a certain degree of disturbance, it may be, of his gums, it may be of his digestive functions. But, with a careful supervision, with moderation and caution in your vigor, you will almost always succeed in accustoming him to accept and tolerate this plan of treatment. †

---

* I do not speak here of mercurial frictions, and for cause. Frictions certainly constitute an excellent mode of treatment, and would, in point of fact, be very useful. But in this contingency they are almost always inapplicable, for it is necessary to have regard to the exigencies of each particular case. How could a young husband accept our prescription to pomade and bundle himself up every night, and then present himself in such attire in the conjugal bed ? Such a plan of medication would little favor the concealment of a condition which the patient has a special interest in concealing.

† I do not say, be it understood, that a treatment of this kind should be applied

As an example of this kind, I will cite the case of a patient that I treated in this way, five years ago, for a syphilis breaking out the *tenth day after his marriage.* This young man (whose history is exactly identical, as regards peculiarity of début, with one of our preceding cases)—this young man, I say, a fortnight before his nuptials, in compliance with the traditional ceremony called *l'enterrement de la vie de garçon,* passed the night with a former mistress whom he thought *safe.* He thus acquired the pox, which, after an incubation of twenty-five days, manifested itself by a chancre on the glans. The situation was then most critical. I employed the method which I have just mentioned, and I have the satisfaction of saying that it was crowned with complete success. The secondary period remained almost entirely quiet; all risk of contagion was averted; all was saved, "even to appearances," following the expression of my patient. And yet it was not without difficulty that I prevailed upon him to undergo this severe treatment. Many times he kicked against my prescriptions, against what he called my "horse-treatment," but which I myself would, in more exact terms, characterize as "a treatment for the use of syphilitic husbands who do not wish to infect their wives."

The first point in the situation which occupies us settled, let us now proceed to a consideration of the second, which relates to the *dangers of contagion* incurred by the wife. These dangers you will recognize from what precedes. They are of two orders:

1. Dangers of direct contamination from the transmission of a contagious accident from the husband.

in all cases; but I say—which is quite different—that, *in cases where it can be applied,* it ought to be made use of. It constitutes, in effect, the surest means of preventing the contagious manifestations of the secondary period.

2. Dangers of indirect contamination, resulting from a pregnancy (syphilis by conception).

Now, it is a question of the preservation of the young wife from these two perils. With this object in view, what must be done?

As regards the dangers of direct contagion, our rôle, our duty, is all traced out: it is to warn the husband upon this particular point in a manner the most explicit, the most complete; it is to frighten him a little as to the risks incurred by his wife. A little scare would not be harmful, in order to render him more cautious and more prudent.

Inform the husband, then, exactly of the dangers of such contagion—dangers which he can not be aware of, at least sufficiently. Do not restrict yourself to simply saying to him (as is generally done), that he may be contagious, and that he ought to abstain from all intercourse with his wife, in case he should be affected with any syphilitic symptoms. That is too vague. Press the matter home to him; do not fear to enter into details, for it is worth the pains, and convince him thoroughly of this— that, in his state of disease, every sore, every erosion, every lesion, excoriative or humid, contains, or may contain, a germ of contagion; that however slight, however insignificant, however innocent any lesion whatever may appear to him, that lesion is no less dangerous on this account; that it matters not, besides, what may be the situation of a lesion, as far as its contagiousness is concerned; that there may be contagious lesions in the mouth as well as upon the genital organs, etc.* Then

---

* A widespread opinion among the laity is that syphilitic contagion can only be transmitted by the genital organs. For them, the idea of syphilis implies that of a genital contamination. This is an error which it is important to correct on

you will add, in conclusion, " *That which you have* imposes upon you the express, absolute obligation to abstain from all intercourse, from all contact with your wife, for there may result from it the worst of contagions to her."

And, gentlemen, you are not only authorized to say this, but it is your duty to say it, and in these terms ; for such statements are in perfect harmony with the facts of science relative to the usual mode of contagion in marriage.

Do you know what clinical observation actually teaches upon this subject? My notes are absolutely explicit in this regard, and permit me to affirm these two propositions :

1. In the enormous majority of cases, *the syphilitic contagions transmitted in marriage from husband to wife are derived from secondary forms of accidents.*

2. *These contagions are almost invariably derived from secondary accidents of a superficial, erosive, or exulcerative, at most, papulo-erosive form*—that is to say, from accidents essentially *benign* in appearance—almost insignificant, by reason of their seeming benignity, without importance, upon the whole, and quite susceptible of being misunderstood as regards their nature, or of even passing unperceived. And it is evident that this twofold proposition results from the essential nature of these conditions.

For, on the one hand, syphilis is infinitely more dangerous in its secondary stage than at any other period, by reason of the extreme multiplicity and the possible dissemination of its accidents ; and, on the other hand, the contagions which take place in marriage can only be the

every occasion, and more especially still in the class of situations which we are now considering.

result of lesions so unimportant that the husband, conscious of his condition, disregards them, or fails to recognize their existence. A husband, in fact, does not infect his wife as prostitutes are infected, who, from interest or indifference, indulge in coition, whatever may be the condition of their health ; a husband never transmits syphilis to his wife, except through ignorance or inadvertence. Then, he transmits it to her only through the medium of lesions so minute, so benign, that he has not suspected their true nature, or may not even have been conscious of their existence.

I have often said, and I shall not cease to repeat it, *the slightest accidents of the secondary period are the most dangerous as regards contagion.* And they are the most dangerous by reason even of their apparent benignity. They seem to be so trifling, they have an appearance so innocent, that one disregards them, that one does not suspect their nature ; and, consequently, one is so much the more liable to communicate them. Let us add, moreover, that they may easily pass entirely unperceived.

The small secondary erosions of the lips, of the tongue, of the penis, are the most usual sources of contagion in marriage. Recall, as examples of this, two cases which I cited to you at the beginning of this exposition. In one, the contagion was transmitted by secondary erosions on the glans, which had been taken for herpes ; in the other, it was the result of minute erosions of the lips—erosions scarcely desquamative, and altogether comparable to those epithelial exfoliations occasioned by the abuse of tobacco. Well, to these two cases I could add at least fifty others, all testifying to the same effect. Almost invariably, then, it is by these minute lesions, by these *simply erosive sec-*

ondary syphilides, that syphilis passes from the husband to the wife.

This is so true, that patients as attentive as possible to the state of their health, as conscientious observers of themselves as could be imagined, suffer contagions of this kind to happen. Physicians, even, thoroughly competent observers, have not escaped this danger in their own families. The following case is an example, which I consider worthy, in every respect, of being cited to you: A most distinguished physician, one of those men who do honor to our profession as much by their personal character as by their talent, contracted syphilis in the exercise of his art. Being married, he immediately forewarned his wife, and watched himself with the most scrupulous care. Each day, night and morning, he examined himself with the greatest attention. And, nevertheless, in spite of all his vigilance, he finally infected his wife. Let us hear him recount his misfortune, in a letter which he did me the honor to address me upon this subject: "One morning last year, upon awakening, I was astounded to observe, in the furrow of the glans, a small spot scarcely apparent, of the size of a lentil, dry in almost its whole extent, the center slightly excoriated, in a point of surface comparable to the head of a pin. I was astounded, because the night immediately preceding this discovery I had had connection with my wife. And, nevertheless, I had examined myself, as was my custom, the evening before. . . . Now, it was undoubtedly this miserable pimple, this insignificant lesion, which infected my poor wife. For, after the classic delay—that is to say, three weeks later—she commenced to feel a hardness upon the vulva, which soon developed into a chancre. Let not my example be lost. Let it be of advantage to you, my dear

friend, who are occupied with such special studies, in
order to tell those who will hear you how contagion
may be produced in marriage, in order to convince them
that this contagion may be effected by a lesion the most
slight, the most inoffensive—so inoffensive, so slight, as to
escape the suspicious eye of an honest husband and of an
attentive and watchful physician."

There is nothing to add, after this sad and very instructive example.

This is not all, gentlemen. A third point claims our
attention. You have not lost sight of the situation, the
study of which we are pursuing. A married man has
come to ask your counsel for the accidents of syphilis.
In the first place, you have prescribed a treatment for
him. In the second place, you have put him on his
guard against the dangers incurred by his wife from direct
contagion. Your task is not finished ; there remains the
danger of a pregnancy ; and this pregnancy, occurring
under such conditions, would be the occasion of a double
misfortune, viz. : of a misfortune which concerns *the
mother*, in exposing her to the danger of receiving the
syphilis from her child ; of a misfortune which concerns
the *child*, which would be subjected to all the risks of
syphilitic heredity. Now, upon you devolves the duty
of preventing this double disaster.

It is to be assumed that your patient has no idea, or
that he has a very incomplete understanding, of the pernicious results which may succeed a pregnancy occurring
in these conditions. It is your duty, then, to enlighten
him upon this point, and to enlighten him *in extenso*,
clearly, comprehensively, so that, with a full knowledge
of the situation, he shall know how to regulate his conduct.

Consequently, in order to fulfill this last indication, you will, if you trust to me, continue your advice with special reference to three points, as follows:

"And, especially, sir, under the present circumstances, there must be no child. Guard well against pregnancy; avoid at any price your wife becoming *enceinte!*—for, on the one hand, the infant which she would conceive by you might either inherit your disease, or, more likely still, die before being born; and, on the other hand, your wife might be infected by her infant—that is to say, might receive from this infant the pox which it would have inherited from you. Then, you understand me well, you must arrange not to have a child."

And you are at liberty, gentlemen, to insist, if need be, and to add a complement of instructions, as you may judge necessary, according to the attitude of your patient; you are at liberty, as M. Diday has very properly said, "to make yourself teacher even as to the most minute details—teacher always decent, but sufficiently clear."

Such, gentlemen, is the first of the situations which syphilis creates in marriage.

However complex and delicate this situation may be in more than one point, it is, nevertheless, the simplest of all as regards the medical indications which it comprehends.

Let us pursue our study, then, expecting difficulties much more serious.

CHAPTER XIII.

A SECOND order of cases is presented, as follows : A man, recently married, has been reattacked with specific accidents, resulting from a syphilis incompletely treated during his bachelor life. His wife remains uncontaminated, but she is *enceinte*. And, justly alarmed, this man comes to request your advice, submitting to you this double question :

"1. What ought I to do for myself ?

"2. Is there anything to be done for my wife and the infant which she carries in her womb ? "

This second situation is much more complex than the one we have previously studied, since it embraces all the difficulties of the first, with the grave complication of a pregnancy superadded.

What is the physician's duty in such a case ?

I. As far as the husband is concerned, there is no embarrassment. Our rôle is exactly what is was in the first order of cases previously considered, and we have nothing to do but to prescribe a treatment appropriate to the character of the existing accidents, and to put our patient on guard, by properly given advice, against the possibility of a contagion which, in this case, would have results doubly unfortunate, since a wife and a child are at the same time interested.*

II. But it is in that which concerns the wife and child that you will encounter the real practical difficulties. It is evident that both are threatened. In the first place, this young mother may be contaminated by this offspring of a syphilitic father, whose syphilis is still sufficiently active to manifest itself by present accidents. Then arise all the dangers of syphilis by conception. On the other hand, this infant is liable, from its hereditary dangers, either to be born syphilitic, or (which is most commonly the case) to die before being born.

Now, this question presents itself: May we not hope to ward off these eminently grave results by anticipating them—that is to say, by administering specific treatment to the mother prophylactically? Is not the attenuating and corrective influence of this treatment demonstrated in cases which, if not identical, are at least analogous? Have we not seen, for example, the opportune administration of anti-syphilitic medication cut short a series of successive abortions, resulting from paternal syphilitic influence, and conduct a pregnancy to full term? The intervention of preventive treatment has here, then, at least a rational indication.

But, on the other hand, are we justified in this pre-

---

* In reference to this last point, I have thought it useless to remind you that a contagion transmitted in the course of a pregnancy involves a twofold order of dangers, viz.: 1. Dangers relative to the mother; these are only too evident. 2. Dangers relative *to the infant*. It may happen that the invasion of syphilis in the course of pregnancy may determine either an *abortion* or a *premature accouchement*. It may happen, also, that the *infection of the mother may be transmitted to the fœtus*, with all the serious consequences of a congenital diathesis. Facts of this kind are so common that it will suffice, I think, to simply announce them, without bringing new proofs to their support. Many examples, moreover, will be incidentally furnished in the statistics which may be found at the end of this volume (vide "Illustrative Cases," Note III).

ventive intervention? What do we actually know of the state of the infant? whence do we derive our fears for the mother? Without doubt, this infant runs risks of paternal hereditary infection: that is incontestable. But we have previously shown that syphilitic heredity has nothing inevitable in it, especially when derived from the father. It may be that the infant has received nothing from its father; and, in this case, the mother has nothing to fear from the infant. It may be, then, that our intervention would be *without object.*

In this uncertainty, what is to be done? Is it necessary to resolve upon a treatment directed, perhaps, against illusory dangers? Or, indeed, should we trust to expectation, and "run the chances," as it is commonly termed? This is a grave question as regards results; and a question, unfortunately, still undecided in the present state of science.

To no purpose will you interrogate your books upon this subject; vainly will you search for a precise, categorical solution. In the greater number of our classical, special treatises, the problem is not even stated. And if you consult professional opinion, as I have done, you will find it singularly hesitating upon this subject. For my part, I have tried this experiment: I instituted a sort of inquiry upon this matter, and consulted a great number of physicians, and I arrived at this result: that certain of our *confrères* express themselves resolutely in favor of preventive treatment; that, to others, this practice is repugnant; while the greater number have no settled ideas, and remain undecided, wavering between two contrary opinions.

I wished to have, in order to communicate it to you, the opinion of an illustrious master—of the man who is

11

the most often found grappling with difficulties of this kind, and whose vast experience is so valuable to consult. I went, therefore, a few days since, to confer with M. Ricord upon this special subject, and I found him, also, hesitating and uncertain. "It would be impossible for me," he replied, "to give a categorical solution of the grave question of which you speak, and one which has deeply interested me for a long time. However, from my own experience, I have been led to believe that the part of inaction, of expectation, is altogether the most judicious in this matter. . . . Whatever may be my desire to save a compromised future, it is repugnant to me to act at hazard, to try a campaign of adventure. I am loath to condemn to a mercurial treatment a young woman who has at present nothing syphilitic; who may, indeed, both she and her infant, have escaped the pox, and whom, besides, a treatment would not, perhaps, save from the pox if she had already received it. . . . Still, I do not condemn, I have not the right to condemn, those who think otherwise, who base their practice upon an intention certainly rational in thus taking a salutary precaution. . . . It is for further experience to decide. But, at present, I confess that my preferences are for the expectant doctrine; and, should a case of this kind present itself to me to-day, I should remain inactive, rather than act at random."

Such is, likewise, the rule of conduct which I should follow, for my part—still, without having, I confess, reasons clinically sufficient to justify it.

Definitely, then, you see, gentlemen, the question remains undecided. It is not because it is still new, but it is so delicate, so difficult to appreciate in a categorical manner, that there is no occasion to be astonished at the absence of a precise solution. See how the clinical observations, which could alone serve to decide the question, are really susceptible here of opposite interpretations. Paternal heredity is not inevitable, as I have many times said to you. So that the children of syphilitic fathers are sometimes born living and healthy ; sometimes they die *in utero*, or are born syphilitic. Likewise, they sometimes leave their mothers uncontaminated, and sometimes they react upon them, communicating syphilis to them.

Now, suppose that in a case of the kind we are considering we try the intervention of treatment. The wife is confined at term of a healthy child, which remains healthy. Should we be authorized in attributing this happy result to the treatment? But we are immediately told that "without treatment the thing occurs just the same," and, in support of this statement, a certain number of well-authenticated facts may be invoked. In order to judge this question conclusively, it would be necessary to make use of a considerable number of observations of this kind, drawing a parallel upon a very large scale between the results of therapeutic intervention and those of expectation. Then, before the imposing figures of such statistics every one would be forced to bow. Unfortunately, statistics of such convincing importance are still wanting to us, and I am forced to repeat, in conclusion, that the problem remains simply stated, without any possibility of a solution being assigned it at present.

I shall make, nevertheless, one reservation. There are, in my opinion, *particular* cases where the doctrine of expectation ought to be abandoned, and give place to an active intervention. What are the cases to which I refer? I will specify by an example : a healthy, well-developed woman, married to a syphilitic man, has had *several miscarriages* in succession, and that without cause, without reason. You examine her most carefully, and you find no other plausible explanation than the syphilis of the husband.

Now, this woman again becomes *enceinte.* Anxious as to the result, she comes to consult you, or (what is more common) one of her family—her husband, for example— comes to consult you on her account. In these circumstances, shall you remain inactive? No, certainly not; for, on the one hand, you know, from the experience of the past, how expectancy would again result, at least, according to every probability; and, on the other hand, you have at your disposition a treatment which, directed against the probable cause of these successive miscarriages, may weaken and correct this cause. Why not make use of this resource? Why not have recourse to this treatment? Here, at least, is a chance to be gained; and this chance you have not the right, it seems to me, to withhold from your patient.*

For my part, in such conditions, I do not hesitate to prescribe specific medication as the sole means capable of parrying the danger which threatens the child, and of conducting the pregnancy to term. I do not hesitate to prescribe mercury. I consider myself justified by the needs of the case in disguising it from the young wife under a respectable pseudonym, in concert with the husband. And, if I am not mistaken, I believe I can claim that this practice has often furnished me with real, incontestable successes.

---

* Professor Depaul has developed the same opinion in his *Clinical Lectures.* According to him, "After a series of miscarriages, for which we can not find a cause, the physician is justified in prescribing, empirically, a specific medication— a medication, moreover, inoffensive when it is properly administered."

CHAPTER XIV.

THIRD situation, which is, unfortunately, the most common. *A syphilitic subject married has infected his wife.*

Summoned in such conditions, what have you to do? What medical indications present themselves to be fulfilled?

"The matter is most simple," you will perhaps say to me; "we have here two patients; well, we will treat these two patients." Without doubt; but that does not terminate, ought not to terminate, your rôle, which, in reality, is much more complex than you would at first suppose. You are, let us assume, in actual practice, and you are about to experience the *difficulties of practice,* which one can not fully appreciate unless he has served a personal apprenticeship.

Let us be explicit.

I. In regard to the husband there is no embarrassment. As far as he is concerned, there is nothing to do but this :

1. To prescribe for him a treatment.

2. To intimate to him, in the most direct and forcible manner, *the interdiction of paternity.*

You know well, gentlemen, how a pregnancy results when both parents are contaminated ; especially when the maternal syphilis is recent, and has not yet been subjected to the depurative effect of treatment. A pregnancy in such a condition is a disaster. Then, it will be your duty to enlighten your patient upon this point, and, in order to leave him in no doubt, you should address him in the following language : "In the present circumstances, with the disease with which your wife and yourself are at the same time affected, a pregnancy would be the worst misfortune which could befall you. For, one of two things would happen : either your child would die before being born ; or it would come into the world with the pox, and you may imagine the effect of this for yourself, for your wife, for your two families, for others, etc. ; without taking into account that the poor creature could not long survive, despite all care. Then, in your own interest, as in the interest of all, avoid, at any price, the possibility of a pregnancy until further orders." Such is the urgent advice to give—more easy to give than to be followed, it appears, as we shall have proof in a moment.

II. Here, then, is the situation regulated, so far as relates to the husband. But there remains the wife. And it is in relation to her that there is about to begin for us a most delicate situation—by so much the more delicate, as it will be necessary to combine with our usual prerogatives the rôle of the tactician and the diplomate.

And, in reality, in the large majority of cases, these things happen in such a way that the wife is entirely ignorant of the disease with which she is attacked, and it is your moral duty to deceive her in this matter, by dissimulating the real name and nature of her affection. Why ? Because nine times out of ten, at least, the situa-

tion involves you with it, as follows : The young husband
who has infected his wife rushes distracted to his physician, and thus begins the interview : "Doctor, a great misfortune has befallen me. I had the pox. I made the
mistake of marrying before being completely cured. I
have given the pox to my wife. I come to ask you to
treat her. But, above all, I beg this of you, I ask it of
you in the name of all you hold most dear : do not disclose the truth to your future patient ; keep her for ever
ignorant of the name and nature of her disease. For, if
she knew this, I should be ruined ; it would be the death-
blow to her affection and esteem for me. And, if she
should tell her family, imagine the result ! Then promise
me your best care for her, and, at the same time, perfect
discretion—an absolute silence as to the nature of her
condition."

Could you, gentlemen, excuse yourselves under such
circumstances, and refuse the double service which is thus
requested of you ? Certainly not.

So that, at the first step, you are here involved in a
singular predicament—that of a physician treating a patient, with the obligation to conceal from her the disease
for which he treats her. A singular situation, I have said,
but a situation to be accepted, since it has nothing incompatible with our professional dignity ; for, after all, we
are not responsible for this situation ; we only submit to
it, and we submit to it from a motive essentially moral
and beneficent, viz., with a view of concealing a culpable
action, and, following the consecrated expression, of preserving the peace of a family.

On the other hand, gentlemen, do not misapprehend
the difficulties, altogether special, of the mission which
you will have accepted under these circumstances. To

treat a woman with syphilis (and to treat her a long time, as is necessitated by the nature of this disease), *without this woman ever divining or suspecting the truth*, is a task which a diplomate might undertake, but for which a physician is poorly prepared. And, indeed, it will be necessary for you to do certain things to which you have not been accustomed, viz., to manœuvre in a line of perpetual dissimulation—to satisfy *ex abrupto* a hundred questions with which your patient will besiege you : "Now, doctor, what is it I have? What are you treating me for? How did I take this disease? How does it happen that I have the same symptoms as my husband?" etc.—to confer upon the different morbid manifestations which arise different pseudonyms, which must be irreproachable, acceptable, probable—to mask under imaginary names the specific remedies which you will have to prescribe ; and, in all this, never to hesitate or evade for an instant, never to betray yourself.

Now, this rôle—as you will appreciate only too fully in practice—comprises more than one difficulty ; it demands an assurance, an *aplomb*, an address, which can only be acquired by practice. In brief, it is less easy, believe me, than one would imagine *a priori* to manœuvre upon such a field ; and more than one able tactician has succumbed in this contest with feminine acuteness. Consider yourselves warned, then, gentlemen, and, when you engage, or, rather, when one engages you, in an undertaking of this sort, do not lose sight of the fact that you will meet your match.

And yet, be assured, the women that we pretend to impose upon in this way are very far from being always the dupes of the stratagem. In reality, we deceive them much less often and less completely than we ourselves think, and, especially, than their husbands think. Many times, for my part alone, I have perceived that certain of my patients, whom I thought I had misled as to the nature of their disease, knew perfectly well what was the matter with them. Only, before myself, as before their husbands, they accepted—because they chose to accept it—the rôle of deluded wives. Some of them, moreover, after a certain time, place the physician at ease by making him understand that they are aware of the situation. "Now, do not give yourself so much trouble," one of my patients said to me one day, "to persuade me that I have a disease other than that for which you are treating me. I have for a long time understood the nature and the wherefore of my disease. Only, so far as my husband is concerned, I shall always remain ignorant, for *my dignity obliges me to ignore that which I could not pardon.*" Another—a woman of intelligence— you will pardon me the anecdote—seemed absolutely confiding in my imaginative diagnostics, until one day she disabused my mind by the following little speech: "I am very much obliged to you, dear doctor, for all the trouble which you have for so long a time taken to dissemble the disease with which I am affected; and you might have succeeded, perhaps, had it not been for my husband and M. Littré; but my husband guarded too preciously your prescriptions not to inspire me with a desire to read them; and I have read them, as you may believe; and you had forgotten to make a recommendation to M. Littré not to indicate in his dictionary the synonym of your fallacious word hydrargyrum." These things happen thus quite frequently, and it is well to be acquainted with them for the demands of practice.

There is another point of capital importance. If it be

difficult, as we have just seen, to treat a woman for syphilis without arousing her suspicions, it is much more difficult still to treat her as it would be desirable that she were, as you would wish that she were treated—that is to say, in an energetic, prolonged, and sufficient manner.

I will explain myself, and I shall not hesitate to insist upon this point, for a considerable interest, which merits all our solicitude, is attached to it.

You know that one cures the pox, or rather that one definitely imposes silence upon the manifestations of the pox, only at the price of a long, very long, treatment, requiring, at the minimum, several years. You know, in addition, that this treatment, in order to be efficacious, requires a particular direction—that it ought to be by turns interrupted, recommenced, stopped altogether, resumed in various forms. All this demands much time and patience, a medical supervision, if not constant, at least prolonged. In a word, the pox is a chronic disease, which one gets rid of only by a chronic treatment.

Now, judge whether, in this special case, such a treatment would be easily applicable.

In the first place, how will you force the acceptance of a treatment of this kind upon a woman who is ignorant of the nature of the disease with which she is affected, to whom you have not the right to explain what the disease is, what its multiple consequences and its dangers for the future may be, etc. ?

And more, how make a woman accept this treatment who is constantly deceived as to her condition, to whom her husband, by way of consolation, or in extenuation of a fault still unavowed, does not cease to repeat to her every day that " what she has is nothing," that "it will soon pass away," etc. ?

Note, moreover, that the said husband, at a certain stage of the malady, as soon as its evident manifestations have disappeared, as soon as the syphilis no longer exposes itself by external symptoms, becomes for you an auxiliary less than ardent. As at first he was most zealous in his efforts to obtain from you an active medication, and to have you supervise its application ; so, later on, you will find him lukewarm, when, the ostensible accidents having been effaced, he no longer insists upon a preventive treatment. You were "a savior," you were welcome in his house a few months ago. But now "that all is finished," "that there is nothing more the matter," your presence with his wife, your visits, your prescriptions, your treatment, "which, without doubt, have done well, but which should have the merit of being less prolonged," all this becomes for him a source of vexation, of irritation, of disquietude, by renewing disagreeable souvenirs, by prolonging a difficult situation, naturally calculated to excite suspicion. In brief, to speak clearly, this husband longs for nothing more than to be disembarrassed of you (the word is strictly exact), and your disappearance from the scene will be a veritable deliverance for him. Hence this lamentable consequence, viz., that *every married woman, contracting the syphilis in the conditions which we are now considering, will never be otherwise than insufficiently, very incompletely treated, and will, on that account, remain exposed to the most serious dangers in the future.*

Such, gentlemen, is the invariable history of women who have been infected by their husbands. In the beginning of their disease, these women have always been treated some little (exception made of some, who, thanks to the selfishness of their husbands, have received no treatment whatever). They have been "whitewashed," permit me the common but consecrated expression. Then the physician is hastened into relinquishing a treatment which might have awakened suspicions, and become compromising for the husband. As soon as possible the physician is dismissed, and *things rest there.* What happens then? The syphilis, be it well understood, does not relinquish its hold upon these unfortunate women, despite their quality of married women, of honest women; and ten, fifteen, twenty years later it is manifested upon them by accidents of diverse forms, more or less severe, very serious sometimes, even mortal.

Add to this another consideration still more aggravating: that, occurring in married women, who appear guaranteed from syphilis by an entire past of irreproachable morality and of thorough respectability, these specific accidents of the tertiary period run the risk very often of remaining unrecognized, and consequently they are not submitted to the sole treatment which is appropriate to them, and they have every chance, on this account, of terminating in the most disastrous results. A diagnostic error as to the nature of these accidents is, in such a case, more than easy to commit. In the first place the physician, by reason even of the character of the person—I mean by reason even of the presumed antecedents of his patient—does not dream of syphilis; he is far from suspecting syphilis in the virtuous, respectable, venerated surroundings where he finds himself called. Can he suspect it, moreover, when he receives from his patient no acknowledgment, no indication—for the excellent reason that this woman can not reveal a disease of which she has always been ignorant? On the other hand, he is hardly more enlightened—at least, ordinarily—by the husband,

by no means eager to revive a compromising past, little disposed to confidences which he judges absolutely useless,
etc. So that, as an almost general rule, a correct diagnosis
is not formed, at least when (which is the exceptional
case) it does not declare itself by the objective character
of the lesions. And I leave you to judge of the consequences of a mistake in conditions of this kind—that is
to say, in presence of lesions so serious as those of tertiary syphilis.* I insist, and I repeat, that *among women
nothing is so frequent in practice as the tertiary accidents of syphilis contracted in marriage.* Observations
of this kind abound and superabound. I count them by
hundreds in my notes of hospital and private practice.
And, for the most part, for the enormous majority, they
relate, I repeat it, to women who have been treated in a
very insufficient way at the début of the disease, who
have been treated only for the time actually necessary
to dissipate the first accidents, in order to save appearances, and exonerate the husband from responsibility
most quickly.

Very far from my purpose, assuredly, is it to pretend
that it is a cold and cowardly calculation of selfishness

---

* One may perhaps say, " But the antecedents of the patient will be known to
her own physician, who, in consequence, will not mistake the nature of any accidents which may afterward happen." " Yes," I would reply, "if the physician
called to examine these last accidents is indeed the one who originally treated the
patient. But, in the contrary case—how then ? Now, there are numerous chances
that the contrary case will be frequently met with in practice. It most often
happens—this is the result of experience—that the physician who is called upon to
treat the syphilis of a woman contaminated by her husband is not the regular
physician of this patient. Almost always he is, after a certain time, supplanted by
a *confrère*, and this through the agency of the husband, who is distrustful and
little anxious to keep near his wife the confidant of a compromising past. Then,
for this or some other reason, it frequently happens that the antecedents of the
patient remain unrecognized, when upon these alone depends the establishment of
a correct diagnosis as to the nature of consecutive accidents.

which induces numbers of husbands to sacrifice in this way the future of their wives to the immediate interests of their own dignity ; or, from personal concern, to cover up the fault of which they have been culpable. That would be an unwarrantable, exaggerated, ridiculous accusation. But I can not refrain from remarking that which exists, and from reading in certain facts the condemnation which they carry with them. I can not refrain, for example, from condemning energetically the conduct of those husbands who, in order to avert suspicion, do not scruple to abridge the treatment prescribed for their wives, and who compromise the health of others in order to preserve what they call "their respectability." Neither can I forbear accusing of carelessness, of imprudence, of indifference, etc., such others as, when once the first symptoms have disappeared from their wives, do not further concern themselves with what may follow ; they let things go as they will ; they repose in a security all the more complete, as their own health is not in jeopardy, and thus prepare, with an absolute indifference, dreadful catastrophes for the future. Certainly, examples to cite in justification of the preceding are not wanting to me. The following, among many others, will strengthen your convictions :

A young lady, of high birth, marries a syphilitic man, and is very soon infected by him. Thereupon, great commotion. M. Ricord is immediately summoned, and treats the patient. Everything disappears. They quickly turn their backs on M. Ricord ; they have no further use for him ; they would almost swear never to have known him. Nevertheless, the syphilis retains its potency, and reveals itself by the ill-omened results of three pregnancies, which, one after another, produce still-born infants. Several years pass in peace. Then the young wife is attacked with singular accidents of the nose. She is troubled, without cessation, with a "cold in the head," and the discharge of an abundance of sanious or purulent mucus from the nostrils. Numerous plans of treatment are brought into requisition, but all without benefit. Two seasons of sulphur-waters produce no better results. A physician at this juncture suspects syphilis, and interrogates the patient in this direction, who, in the ignorance of her specific antecedents, quite naturally defends herself by indignant denials. The husband, who is present at this consultation, remains impassible and mute, persuaded that "his wife has been cured of what she formerly had, and that the present accidents have nothing to do with the *little misfortune* of earlier days." Still the nasal lesions continue and become aggravated, until they result, on the one hand, in a frightful ozæna, and, on the other hand, in a perforation of the palate. Then, only, does the husband's conviction become shaken. Then, only, does he consent to recall M. Ricord, who immediately recognizes the specific nature of the disease. I am summoned in turn, and have only the easy rôle of confirming both the diagnosis and the treatment of my illustrious master. But at this time the lesions have become such that the whole nasal bony structure is necrosed. Three entire years this unhappy woman is condemned to absolute seclusion, on account of the insupportable odor which she spreads around her, and which every imaginable medication succeeds only in imperfectly correcting. She is cured only after the expulsion of numerous sequestra and the complete loss of the palate.

Likewise, I have among my notes numbers of other observations relative to women who, after having acquired syphilis from their husbands at the beginning of marriage, have only been subjected to short treatments, and have afterward experienced the most severe tertiary accidents —this one, for example, a phagedenic syphilide which invaded the face, and disfigured it most horribly; that one the loss of the nose; another a rectal stricture, which had to be operated upon, with failure to relieve the patient; another a cirrhosis which, misunderstood as to its nature, carried her off rapidly; still another, lesions of the cranial bones and cerebral gummata, which produced, successively, epileptiform attacks, hemiplegia, gradual fading of the intelligence, dementia, and death; etc., etc.

Accidents of this kind are certainly quite common, and I do not give them to you as constituting manifestations peculiar to the order of cases which occupy us at this moment. But certainly, also, it is at the same time both remarkable and distressing that such accidents are *common in marriage.* Again, they are frequently encountered in the highest classes of society, where the multiple conditions of social position, of knowledge, of civilization, of moral training, ought, it would seem, to exclude such shames. Finally, and still more, they impose a responsibility upon those who, the first authors of the evil, were under obligations to do everything to avert its terrible consequences, and who, notwithstanding, have for one reason or another evaded this direct duty.

In such circumstances, gentlemen, a humanitarian rôle is also imposed upon you; and this rôle you have already comprehended in advance, you have already anticipated in the course of the preceding discussion. This rôle is to take care of the woman who is confided to you, and in relation to whom you have until now only been the *accomplice of the husband,* charged with deceiving her as to the nature and possible consequences of her disease; it is to protect the health of this woman in the present and in the future from the neglect of this selfish or indifferent husband; it is, in a word, to employ every means, through your double influence of physician and man, that this woman may have the benefit, like every other patient, of an energetic, prolonged, active, sufficiently preservative treatment.

It goes without saying that in numerous cases this rôle will be rendered easy by the good disposition of the husband, a man of heart, regretting bitterly the misfortune of which he is the cause, and ready to do everything in his power to repair his fault. It will suffice, then, to explain to him the future dangers to which his wife will remain exposed in the absence of a sufficient treatment, in order to obtain from him a *carte blanche* relative to the direction and the duration of your therapeutic intervention.

But, on the other hand, you must expect to encounter other cases where your position will become much more delicate, much more embarrassing; when, for example, you will have to contend with the selfishness, the indifference, the prejudices, the fears, the ignorance, of the interested principal. In such a case, it will be your duty by your tact, by your experience of the human heart, to struggle with difficulties of diverse kinds in the best interests of your patient. It will be your duty to sustain this contest with a persevering ability; and, finally, if you perceive yourself overruled, to energetically remind the husband of the duties which are imposed upon him as upon yourself in such a situation. I do not advise you, assuredly, to make an exposure, to pose yourself as a redresser of wrongs, to undertake the ridiculous rôle of a Don Quixote of married women. But what I say to you is, if you are

12

forced to do it, attack the position from the front, by addressing to the husband language both firm and severe, such as the following, for example: "Without doubt, sir, it may be extremely disagreeable to you that the treatment of your wife needs to be still continued for a long time, but it is not in my power that it should be otherwise. I have consented, in order to save you from a false step, to be your accomplice against your wife, and I have given you my word. But I will not go further, and I throw upon you the whole responsibility of what may follow, in case, by your act, the treatment should not be pursued as it ought to be. Honor and humanity demand that you do for your wife what you would think necessary to do for yourself. Allow me, then, to finish the work begun, and complete a cure which ought to be the object of our common efforts."

You may be sure, gentlemen, that in speaking in this way, in accepting and sustaining the rôle which I have just traced for you, you will fulfill a moral duty of which you have not the right to divest yourselves. Be equally sure that you will accomplish a salutary service in thus snatching from the tertiary grasps of the diathesis the unfortunate women who were not designed for the pox, but whom the selfishness or the carelessness of their husbands would leave easily exposed to drain to the dregs the chalice of the disease.

Still this is not all, and you have not yet exhausted the embarrassments of this special situation.

On the other hand, indeed, you will have to contend with a difficulty of a different order, viz., with the repugnances, with the resistances, which you will encounter from the person interested, on the part even of the patient that you have to treat; and your position in relation to her will demand a certain cleverness, even a certain professional tact, on your part. I explain:

You have cured this woman, I suppose, of her first accidents. All has disappeared; all is going for the better. And now you speak of a new treatment. But why this treatment? Wherefore, and for what indication? you are asked. Admit that your remedies are again accepted this time. But what welcome will you receive when you return to the charge a third, a fourth, a fifth time, etc.? And, nevertheless, necessity imposes this; you must obtain this from your patient; it is the price of her cure. This difficulty—and many others of the same kind—you will only overcome by dint of patience, of professional tact, of practiced cleverness. You will only overcome them by being able to exercise over your patient an authority which will assure her confidence and render her docile to your prescriptions.

Then, in view of the future, do not neglect to establish the situation well at your first visit, and that in a sense the most favorable to the purpose you have in view. Assuredly, you are under obligation to conceal from your patient not only the name of her disease but the special dangers to which she is exposed, and the terrible consequences which may result after an indefinite period. On the other hand, however, do not commit the mistake of declaring to her, as her husband would wish, that "she has nothing," that "her affection is a passing, trifling indisposition, which will not be followed by consequences," etc. On the contrary, as soon as you may judge it prudent and opportune, express clearly the opinion that her present symptoms indicate a veritable disease — a disease which you may, however, disguise under whatever respectable pseudonym seems to you best under the circumstances. Let her understand that this malady will probably be of long duration ; that it may have disagreeable returns, relapses, recurrences. Assure her that the affection to which she is subject ought to be the object of an attentive surveillance ; that it is quite curable, without any doubt, but that, in order to be cured, it will demand prolonged care, etc. In brief, prepare the way and anticipate the objections, the rebellions, that you have to fear in the future, by indicating in advance the long therapeutical intervention which will be necessary. .

I do not maintain by any means that you should frighten your patient. That would be absurd, out of place. But, believe me, you will defeat the end which you have in view by reassuring her too much, by not allowing her a certain degree of vague apprehension which will serve you as a valuable auxiliary. All patients (and women more than others) are so constituted that they will not be treated when they think they have nothing to dread, when they do not experience "a little fear" (permit me the word). Do not overlook this general disposition of the human heart, and utilize it for the interest of your cause—that is to say, for the protection of the patient, whose health is confided to you, under conditions so especially difficult and delicate.

CHAPTER XV.

FOURTH order of cases: *Husband syphilitic; wife syphilitic and enceinte.*

We now come to the fourth and last situation possible. A man is married despite a syphilis not yet extinct; he has infected his young wife; and, in addition, this wife is *enceinte.*

It would be superfluous to say that this situation is more serious than any we have previously studied, the most prolific in dangers, in catastrophes of various kinds, as also in practical difficulties for the physician.

What may, in reality, result from such a state of things?

In the first place, the infant procreated in such conditions—that is to say, the issue of a father and mother both syphilitic—is subject to the gravest prognosis. It is destined, it may be, to die after some months of intra-uterine life; it may be, to come to term still-born or moribund; it may be, to be born with the pox.

Of these three alternatives, let us concede the best—the infant is born *with the syphilis.* Now, this child may be confided to a healthy nurse, and you may be sure that it will transmit the infection to her—that is almost inevitable. This nurse, in her turn, may infect her own child;

may infect her husband, as has been seen so many times. But I spare you the enumeration of the possible *ricochets* of these last contaminations.

Such is, in brief, the perspective offered to us by this fourth order of cases ; it is, then, most certainly a subject of study which demands all our attention. In the first place, I will reassure you by saying : The situation is very critical, very serious, assuredly ; but it is not desperate, either for the present, or *a fortiori* for the future.

As for the future, the outlook is not doubtful. For, treat this syphilitic couple actively, methodically, protractedly, and you will bring it about that they will have, later on, living and healthy children, exempt from every specific phenomenon. And, even as to the present, as regards the existing pregnancy, it is not impossible—it is not *impossible*, I say no more than this—that treatment may avert a complete disaster. In other words, it is not impossible that the mother, submitted to an active treatment during gestation, may give birth to a living and viable child, and one which may even remain free from every specific accident. Thus :

1. Many times I have obtained from specific medication this prime and inestimable success—the *prevention of abortion*, and the bringing of the pregnancy to full term. The child born under these conditions does not, it is true, escape syphilis ; but it is born viable, resisting, capable of tolerating specific medication, capable, in a word, of living with the syphilis, and of being cured of this syphilis by subsequent treatment. Examples of this kind are so numerous that I regard any particular citation as superfluous.

2. In such circumstances, I have even seen children born healthy, exempt from every syphilitic symptom.

Example: A young man is married while in the active secondary period, notwithstanding the advice of his physician. Five months later, his wife becomes *enceinte*. In the second month of pregnancy she is brought to me, and I discover upon her specific accidents as manifest, as indubitable as possible. I then treat her energetically, and I pursue the medication with vigor during the entire period of her gestation. I have the happiness, first of all, of bringing the pregnancy to full term. Besides, the child is born healthy, well-nourished, and of almost the average development. It continues to live, and remains exempt from all specific manifestations. I observed it most carefully for about fifteen months, after which I lost sight of it; and I can guarantee that during all this period it did not present the least suspicious phenomenon.

M. Langlebert has related a case almost identical. "Madame X—— married, in November, 1869, one of my patients, whom I had been treating for some months for constitutional syphilis. She immediately became *enceinte*, and must have contracted her husband's disease about the same time. She had scarcely completed the third month of her pregnancy when a confluent roseola covered her body. Blackish crusts disseminated over the hairy scalp, a very pronounced alopecia, a cervical adenopathy, tonsillar mucous patches of ulcerating form— everything seemed to indicate the début of a very severe syphilis that she must inevitably transmit to her infant —if, indeed, this infant should see the light, which then appeared by no means probable. I immediately prescribed for Madame X—— pills of the sublimate; afterward I submitted her to the use of the iodide of potassium, continuing the mercury at the same time. The pregnancy followed its regular course, and toward the end of August, 1870, Madame X—— gave birth to a very small but healthy girl, which she nursed herself, according to my advice. Now, this child has not ceased for a single moment to continue in good health. She has had nothing upon her body—neither spots, nor lumps, nor the least symptom of suspicious appearance. To-day she is a little over two years of age ; she is large, well developed, and is marvelously well. She has, then, escaped the syphilis, and this result she owes to the treatment, which could alone preserve her from the infection which her mother's condition during gestation must otherwise have rendered inevitable.*

Successes of this kind, obtained under conditions so unfavorable, are well adapted, certainly, to encourage the physician, and indicate to him the line of conduct to be followed in like circumstances.

That settled, we now come to the indications to be fulfilled in the class of cases which it remains for us to study. Relatively to the husband, there is nothing more simple— nothing more to do than to prescribe for him the usual treatment for the diathesis. But it is the wife, especially, that claims our solicitude at present. This wife must be treated, and treated with so much more of care, of method, of attention, of vigilance, as she represents two patients, if I may thus express it, two beings to be preserved. And, indeed, it is a question of the mother, first of all ; but it is no less a question of the infant which she carries in her womb, of the infant more endangered than she, and which we can only reach, only protect, through her.

To *treat the mother*, then, is the capital indication to be fulfilled.

---

* *La syphilis dans ses rapports avec le mariage*, p. 237.

Well, gentlemen, this indication, so simple, so rational, and, moreover, so completely authorized by experience, you must not regard as accepted by all. It has its opponents. It has excited objections ; it has given rise to controversies which have recently again agitated one of our learned societies.

It has been said : "What! This pregnant woman you are going to treat, and to treat how? You will prescribe for her *mercury?* But this mercury, do you not fear that it may be prejudicial to her in several respects? Will it not, in the first place, increase and complicate the gastric troubles of pregnancy? Will it not add its own special anæmiating action—to the anæmia, the hydroæmia of pregnancy? And especially, a capital danger, does it not involve the risk of producing abortion? For, every day, we see abortion produced in syphilitic women treated by mercury," etc.

To all that, gentlemen, our reply will be as direct, as categorical as possible. Yes, without doubt, we will say, mercury *maladministered* is open to such objections. Yes, without doubt, with mercury given in certain forms or in certain doses we may produce the accidents alluded to — that is to say, aggravate gastric troubles, increase anæmia, and even favor or determine abortion.* But that is not the question, and we have here nothing to do with the toxic or industrial use of mercury. What is alone relevant to our present inquiry is the safe and prudent administration of mercury as *a remedy;* it is a *mercurial treatment* appropriate to the forces and to the special conditions of the patient. Now, a treatment of this kind,

---

* Vide Ad. Lizé, *Influence de l'intoxication mercurielle lente sur le produit de la conception* (*Union Médicale*, 1862, t. i, p. 106). H. Hallopeau, *Du mercure, action physiologique et thérapeutique* (*Thèses d'agrégation*, Paris, 1878).

methodically instituted and supervised, will not only remain innocent of the dangers which have been ascribed to it, but it will, moreover, constitute the best and the surest means at our disposition of bringing the pregnancy to term, and of preserving the fœtus in such a case.

Now, let us enter into details, and discuss, point by point, the various objections which precede. The matter is worth the trouble, since the existence of the infant is here in jeopardy, and depends upon medical intervention or non-intervention:

I. In the first place, as regards the gastric troubles, experience demonstrates that we can easily avoid them. We should guard against administering to our patients the sublimate, the biniodide, the syrup of Gibert, or all other analogous preparations, which are badly tolerated by women in general, and especially by women in a state of pregnancy. We should be careful to prescribe other mercurial combinations which are not apt to disturb the digestive functions to the same degree. We should prescribe, for example, the proto-iodide, a milder remedy, which, in a medium dose, from five to eight, or even ten, centigrammes daily, is ordinarily well tolerated by the stomach. Here we are accustomed to administer the proto-iodide every day to our syphilitic women in a state of pregnancy, and, nine times out of ten, we see them remain undisturbed in their gastro-intestinal functions. Should it determine, exceptionally, some *malaise*, some gastric or intestinal disorder, we almost always succeed in causing it to be tolerated by some expedient or another; it may be by giving it before or during the meals; it may be by associating with it a small dose of opium; it may be by prescribing some digestive adjuvant, such as wine of quinquina, wine of gentian, coffee, etc. If, however, the stomach still shows itself rebellious to this remedy, if gastric or intestinal troubles continue to be occasioned by it, there always remains a resource by which the patient may enjoy the benefit of a mercurial influence without injury to the digestive functions. This resource you have anticipated in advance : it is the resort to *mercurial frictions*, a mode of treatment whose energetic action needs no longer to be demonstrated in a general manner, and which has been highly extolled in a special manner by some physicians as peculiarly adapted to syphilitic women during the state of gestation.

II. The second objection is altogether theoretical. Never, for my part, have I seen the anæmia of pregnancy increased under the influence of a mercurial treatment intelligently conducted. And as to the special anæmia of syphilis, it is now well demonstrated that it has its true remedy in mercury. It has been said, with all reason, that from the point of view of the phenomena of specific anæmia, "*le mercure est le fer de la vérole.*"

III. Finally, it is absolutely false that mercury promotes abortion in syphilis, as certain physicians contend. From the fact that it is not rare to see syphilitic women abort during, or at the termination of, a mercurial course of treatment, it has been inferred that the abortion is due to the mercury in such case. This induction is, in truth, very illegitimate ; it is even, I would say, devoid of all foundation, for it disregards a factor more than essential in this matter, viz., the disease itself, syphilis. *It attributes to the treatment that which is the result of the disease.* It is needless to remind you again that syphilis causes a most powerful predisposition to abortion ; there are few morbid states that may be comparable to it from this point of view, and that furnish so considerable a contingent to the sum total of abortions.* So that, when a syphilitic woman, submitted to a mercurial treatment, comes to have a miscarriage, one is justified in attributing this miscarriage, not to the influence of mercury, but to the exclusive influence of the specific diathesis.

Do you wish the proof of this? This proof may be found in these two results of clinical experience, viz. :

1. That numbers of syphilitic women abort without ever having taken an atom of mercury. The frequency of this first fact is notorious.

2. That numbers of syphilitic women, who, without treatment, have had a series of miscarriages, succeed in carrying a pregnancy to full term only after having undergone a mercurial treatment. This is a point upon which I have insisted at length in a preceding chapter, and which it will be sufficient, I think, to simply enunciate again without more ample development.

So the opinion which considers mercury a cause of abortion in syphilis should not prevail against what I would call clinical evidence—that is to say, against the imposing mass of clinical facts, which, collected from all sides, related by observers exercising their art in the most different quarters, all accord, nevertheless, not only in exculpating mercury from this special accusation, but also in presenting it as the best safeguard that we possess against the abortive tendencies of syphilis. This opinion I energetically condemn ; for my part, I do not hesitate to characterize it as mischievous, in fact, for it has as a logical consequence the privation of syphi-

---

* I may be permitted, in this connection, to refer the reader to a chapter of my *Leçons sur la syphilis étudiée plus particulièrement chez la femme*, p. 955. Vide also " Illustrative Cases," Note I.

litic women in the state of gestation from the benefit of a potent remedy, which is equivalent to the surest means of condemning them to the probable chances of abortion.

For the rest, there is to-day, with very rare exceptions, an agreement upon this question; and, without insisting further, I will summarize, what may be called the present state of science relative to this subject, by formulating the two following propositions :

1. Mercury does not always prevent abortion in syphilitic women, but nothing demonstrates that it has ever contributed to its production, at least when administered in therapeutic, non-excessive, non-toxic doses.

2. It often succeeds in a very evident manner in preventing abortion, in prolonging pregnancy, in conducting it to its normal term.* In practice, then, when we meet

---

* What authorities could I not quote here ! Let me cite at hazard: ". . . . I should be disposed to place syphilis among the most frequent causes of abortion. . . . We may, nevertheless, be quite sure of destroying this cause of abortion as soon as we are able to recognize it. *Mercury properly administered almost always succeeds.* . . . We commonly fear to have recourse to mercury during pregnancy, because we imagine that it may cause abortion. But a large experience has convinced me that this opinion is devoid of all foundation, and that, with prudence, we may administer mercury in doses sufficient to cure all the symptoms of syphilis during the entire period of pregnancy, without any injury to the mother and the child. . . . When a pregnant woman is evidently attacked with syphilis, or when I have strong reasons to believe her infected, I never hesitate to make use of this great remedy. This course has always appeared to me advantageous."—Benjamin Bell, "Treatment of Virulent Gonorrhœa and of Venereal Diseases " (trans. of Brosquillen, t. ii, p. 608).

"Gestation, so far from contraindicating the employment of energetic treatment, demands still more attention and intelligent promptitude. I have seen many more abortions in syphilitic women not treated than in those who, taken in time, have been subjected to a methodic medication."—Ricord, *Traité pratique des maladies vénériennes,* 1838. That which M. Ricord intends by a "methodic medication " is none other than the usual treatment of syphilis with mercury.

"The mercurial treatment is regarded at Lourcine as the *preservative against abortion* " (Coffin, work cited).

"Administered properly, mercury is the most powerful preservative for the with a syphilitic woman in a state of pregnancy, our first care should be to submit her to specific treatment. And if this woman—as, indeed, is the case in the class of situations which we are now studying—is affected with a syphilis still recent, demanding the employment of mercury, we should not hesitate to prescribe mercury. We should prescribe it, without doubt, in moderate doses, appropriate to the patient's strength and gastric tolerance ; but we should prescribe it in an active, sustained, prolonged, veritably efficacious manner, sufficient, in a word, to accomplish the object we have in view.

And this treatment, gentlemen—I do not fear to repeat

---

infant ; and, as M. Vannoni has established (*Il raccogl. med.*, August, 1872), if it does not prove itself a preventive of abortion, it is because it is not given early enough or for a sufficiently long time. There is an urgent necessity of taking pregnancy into consideration in the therapeutics of syphilis, since we have seen a mercurial treatment administered to pregnant women preserve the infants born in the earlier confinements, and allow the disease to destroy those of later pregnancies when treatment was omitted."—Rollet, *Traité des maladies vénériennes,* Paris, 1865.

"The dangers of giving mercurial preparations to *enceinte* women have been much exaggerated ; it is now recognized, on the contrary, that they render *immense service* when syphilis has been the suspected cause of former abortions, for the disease may rest latent in the mother, or attack the fœtus only. It is doubtless in such cases of obscure cause that Young, Beatty, and Russel, in England, have obtained successes of which they refer all the honor to mercury."—Devilliers, article *Avortement* in the *Nouveau dictionnaire de médecine et de chirurgie pratiques,* t. iv, p. 323.

"Without any doubt, the administration of mercury, of the iodide and of other medicaments which contain a toxic principle, when carried to the point of producing a sort of chronic intoxication, is a powerful cause of abortion, and the cases of abortion attributed to mercury are not all from errors of interpretation. . . . But it is no less true—and observation confirms it every day to every mind divested of preconceived ideas—that a treatment during pregnancy by mercury or by any other active agent in order to combat syphilitic symptoms, to destroy the diathesis, or build up the constitution, so far from being a danger, is, on the contrary, an advantage both for the mother and for the infant, if this treatment be directed with prudence, and given in moderate doses."—Jacquemier, article *Avortement* in the *Dictionnaire encyclopédique des sciences médicales,* t. vii, p. 539.

it again in conclusion—we should institute with so much more of care, we should supervise with so much more of method, of solicitude, of vigilance, because it is not a question of only one patient to be cured, but because with this patient and through her there is another existence to be preserved, that of the infant which at this period intimately shares its mother's destinies.

CHAPTER XVI.

WE have just passed in review the four different situations which may be presented, when syphilis has been introduced into a household by a syphilitic husband. And *à propos* of each of these situations, I have endeavored to trace out for you the line of conduct to be followed by the physician, and determine the numerous indications which it is incumbent upon us to fulfill under such circumstances.

Our subject, nevertheless, is not altogether exhausted. An essential and most practical point remains for me to mention to you ; and with it we will terminate our present lecture.

This point relates to a veritable *social* duty (you will see that there is no exaggeration in this word), which is imposed upon the physician in the particular circumstances which we have just been considering—a duty at once manifest and undeniable, and the accomplishment of which is fruitful in useful results, but, nevertheless, a duty often omitted, neglected, violated even, in ordinary practice, to the great detriment of those whom it is our professional mission to protect.

The greater number of our classic treatises remain absolutely silent upon the question which is about to follow.

You will permit me, then, to treat it somewhat in detail, in order to show you its importance and practical diffi- culties.

When syphilis has contaminated a husband and wife there is a great risk, as we have previously shown, of their child being born tainted with syphilis. Now, this infant, supposing it to be syphilitic, evidently carries with it the *dangers of contagion.* That is to say, it is possible that the syphilis with which it is affected may radiate from it to the persons who surround it, who are called upon to take care of it, and who, in various ways, come in contact with it.

Well, then—and this is the point to which I wish to direct your attention—what will happen should this in- fant be confided to a nurse? The answer is easy: almost infallibly this child will infect the nurse. It is thus, then, that syphilis proceeds from the family of the infant and attacks persons outside. First misfortune, first deplora- ble consequence of the situation which now occupies us.

But this is not all. You know, since you have often- times heard me repeat it,* what a singular faculty of expansion, of irradiation, the syphilis of nurslings and nurses presents, which propagates or may propagate it- self by a series of unexpected *ricochets* in such a way as to constitute a source of multiple contaminations. How many times, for example, has it not happened that a syphilitic nursling has infected many persons of its *en- tourage,†* or, indeed, that a nurse infected by a syphilitic

---

* *Nourrices et nourrissons syphilitiques,* Paris, 1878.

† Example of the kind: A nurse infected with syphilis comes into a young family, whose infant is confided to her. She infects this child. The nature of the morbid symptoms remains unrecognized at first, as would almost necessarily be the case, so that no precautions are taken against the possible dangers of such a contam- ination. What happens? The infant, on its part, infects—1. Its mother; 2. Its grand-

infant has transmitted the disease to her own child, to her own husband, to a strange nursling? And how many times, also, has not each one of these new contagions become, in its turn, the origin of other contagions?

The cases in which these *cascades* of contagion are produced, if I may thus speak, primarily originating from the syphilis of a new - born child, abound and superabound. Without exaggeration, they literally swarm in medical literature. I have already cited a great number of them when giving you the history of the *syphilis of nurses and nurslings.* Allow me to briefly recall to you, as types of this kind, the three following cases:

1. A young man affected with syphilis marries prematurely. He immediately infects his wife. A child born of this marriage is confided to a nurse, and infects this nurse. She, in her turn, transmits the syphilis to her

mother; 3 and 4. Two nursery-maids of the family, girls of absolutely irreproachable character. And the young wife transmits the contagion to her husband some months later.

I have often said that *nothing is so dangerous to the persons surrounding it as a syphilitic infant.* The thousand attentions which the new-born requires in its raising—the kisses, the caresses, which are lavished upon it—serve as the origin of easy and frequent contaminations. I have, in my notes, to speak only of cases observed by myself, a dozen cases of contagious of this order. It is thus, for example, that a grandmother, sixty-five years of age, was infected by her little grandson that she fed with a spoon, having taken the pains to carry each spoonful to her mouth before giving it to the child; the virus was certainly transmitted in this case from the lips of the infant to those of the grandmother. Likewise, I now have under my care a young woman who was infected by her child, which had been infected by a nurse. My learned colleague, M. Hillairet, has recounted to me the following case: A young man affected with syphilis marries prematurely, and soon infects his wife. A child born of this couple presents the accidents of hereditary syphilis several weeks later, and infects its nurse. Confided then to its maternal grandparents, it transmits the contagion to both of them by the intermediary of a nursing-bottle. The grandfather and the grandmother had the habit of putting the nursing-bottle to their lips, and that without taking the precaution to wipe it after it came from the mouth of the child. Now, the child being affected with labial syphilides, both were infected in the mouth, and presented an indurated labial chancre, soon followed by general accidents.

own child, in the first place; then, to another nursling;
then, two months later, to her husband.*

2. A syphilitic child, born in an apparently healthy
condition, is confided to a nurse, whom it soon infects.
This nurse, who is nursing another child at the same time,
infects this child, which soon dies. She then takes a
third nursling, which, in its turn, contracts the syphilis
and dies. Another nurse, a friend of the former, having,
through kindness, given the breast to this last child, receives the syphilis from it. She then infects her nursling.

Please remark, gentlemen, this makes five contagions
from this syphilis, and two deaths.

3. Another example, related by one of our most distinguished colleagues—M. le Dr. Dron (of Lyons): A
syphilitic child infects its nurse. This nurse, in order to
empty her breast, suckles three nurslings—all three of
which take syphilis. Each of these children infects its
mother; each of these three mothers infects her husband.†
Count again: ten syphilitic contaminations derived *par
ricochet* from the syphilis of a nursling! And do they
stop here?‡

It is useless to add, moreover, from another point of
view, what you well appreciate, what goes without saying,
viz., that a syphilis derived from such an origin has every
chance of remaining unrecognized, at least during a certain time; and, in consequence of being abandoned to

---

* Vide " Illustrative Cases," Note V.

† Achille Dron, *Mode particulier de transmission de la syphilis au nourrisson par
la nourrice dans l'allaitement*, Lyon, 1870.

‡ Sometimes even (but this is only exceptional, it is true) similar cascades of
contagion have made a still more considerable number of victims. Thus, one has
seen a syphilitic nurse, going into a small village, transmit syphilis to sixteen,
eighteen, twenty-three persons, and become the origin of a small local epidemic
(*vide* Amilcare Ricordi, *Sifilide da allattamento e forme iniziali della sifilide*, Milan,
1863).

itself, left without treatment. Also, that it can not fail, and does not fail, in numbers of cases, to result in veritable catastrophes—it may be, for example, in the death of the nurslings contaminated by their nurses ; it may be in serious accidents developed upon the nurse, or upon other persons who have been the victims of such contagions.*

Now, to return to our subject, gentlemen, it is precisely dangers of this kind which the physician ought to foresee, when he finds himself in a position to prevent them. He recognizes these dangers ; he knows that they will occur if he does not interfere ; it is his business, then, to interfere in order to avert them. And here begins for him an actual duty, which, without exaggeration, I have already characterized as a *social* duty, since it has for aim and for result the protection of the interests of society. This duty, I do not hesitate to say, is *imposed upon the physician*, who would be culpable in neglecting it, in divesting himself of it ; by so much more as, in fulfilling it, he will at the same time satisfy the interests of his patient.

This principle stated, let us come to its application :

To circumscribe the pox in its original bed so as to

---

* Here are examples of the kind, selected from many others :

1. An infant, born of a syphilitic father, is confided to a healthy nurse. It soon presents various syphilitic accidents and infects its nurse. She, in her turn, infects her husband. The husband is affected with an iritis and *loses one eye.* The wife is attacked some years later with a syphilitic paralysis, to which she succumbs (Dr. Delore de Lyon).

2. One of my syphilitic patients marries, in spite of me, and transmits the syphilis to his wife soon after his marriage. A child is born, which (without my knowledge, it is needless to say) is confided to a nurse. This child soon presents numerous symptoms of syphilis, and infects the nurse. This woman, in her turn, infects—1. Her child, which dies in some months ; 2. Her husband ; affected with a severe iritis, *he loses an eye.* A year later, she is delivered of an infant which presents grave accidents of syphilis, and dies at the age of two months.

prevent the spread of its ravages outside, such is the object to be realized. Now, how attain this?

For this there is but one practical means: it is to so arrange that the syphilitic infant, the first origin of the dangers which we are seeking to avert, remains in the family, and is suckled by *its mother*.

It is evident that, if it does not leave the paternal hearth, if it receives its mother's breast, there will be no opportunity of its transmitting to a nurse and to other persons the dreadful contagion of which it bears the germ.

It is, then, to this object that the physician should direct his efforts. He must, by his influence, by his advice, by his moral authority, arrange a situation which will protect the interests of all, and not allow a different arrangement, prejudicial to all, to be organized independently of him. It is necessary, to speak clearly—

1. That the new-born child, the offspring of syphilitic parents, be retained under the roof where it was born, so that he may watch over this child, treat it, if necessary, and suppress as quickly as possible the contagious accidents which may arise.

2. And especially is it necessary that he oppose with all his might this child being confided to a nurse; he must make the family accept, as an absolute, unavoidable necessity, the nourishment of this infant from the maternal bosom.

Let us now come to the practical point. In brief, what is to be done in order to attain this result?

This: At a favorable moment, when the pregnancy of the mother is so far advanced as to permit the hope of an accouchement at term, to address yourself to the husband, and explicitly expose the situation to him, with all the dangers it involves—to say to him that his prospective infant runs the serious risks of hereditary syphilis; to make him comprehend that under these circumstances the child should not be intrusted to a nurse, who would almost inevitably receive the contagion; to unfold to him, without reservation, all the consequences of such a contagion—the just and outspoken recriminations of the nurse, a scandalous exposure, a possible lawsuit, humiliating publicity, etc. ; to conclude, finally, with the absolute obligation imposed upon the mother to nurse her child, the sole moral expedient, proper, and at the same time efficacious, which can retrieve the situation.

"Then," you will add, "arrange it, sir, that your *wife shall nurse this infant.* The interests of all—yours and your infant's—depend upon this. Accept it as an absolute obligation, as an indispensable necessity, considering the circumstances under which you are placed. Therefore, if your wife contemplates nursing it, do not dissuade her from it. And, if she is not disposed to do so, anticipate her objections, and endeavor by all the means which you can bring to bear to modify her resolution. For, from every consideration, I again repeat it to you, it is she and she alone who ought to act as nurse for your infant."

By thus stating the situation, it will be rare, indeed, that the physician does not accomplish the purpose he has in view. Consequently, his object will be realized; the infant, syphilitic, or suspected of syphilis, will remain in the family, and will be nursed by its mother, and thus *obviate danger to others.**

---

* If I had not considered at length this subject in another work, I would insist here upon numerous practical details which I pass in silence. It appears to me indispensable, nevertheless, to add to the foregoing some considerations

Now, this result, gentlemen, do not doubt it, will be a considerable, capital service rendered to public prophylaxis. To be convinced of this, recall what I have elsewhere demonstrated in tracing the history of the syphilis of nurses and nurslings. Recall the frequency of these contagions transmitted by nursing; recall the disastrous, lamentable consequences which result from them, the

relative to a point of especial importance. I will borrow them from my *Leçons sur les nourrices et les nourrissons syphilitiques* (Paris, 1878).

" . . . . Do not hope, gentlemen, that your advice to confide the nursing to the mother will always be accepted without opposition. Without speaking of reasons which do not exist, of reasons based upon pretended conveniences, upon society, or other considerations, it will often be objected that the mother 'is too feeble to nourish it,' that she could not endure the fatigues of nursing without danger to herself, etc. Insist (for it is quite rare that a mother can not, at least for some months, nurse her child)—insist, and say this, 'It may be that the mother can not nourish it as long as an infant ought generally to be nursed, but she should do what is possible, and that is all we ask of her. Let her give it the breast during the first months ; that will enable us to form an opinion, and we will advise after that time. At all events, there is an urgent necessity that the mother nurse it during some months.' And wherefore this, gentlemen ? Wherefor require at least these few months of maternal nourishment ? It is that when infantile syphilis must reveal itself, it does reveal itself, if not always, at least almost always, within the first two or three months. In 158 cases, M. Diday has seen it break out 146 times within this period. Such figures have such a significance that we may dispense with all commentary.

" Then these few months of maternal nursing may serve us as a *criterion of the health of the infant*, and as a *guide for our subsequent action*.

" And, in effect—1. If within this period of time syphilis reveals itself in the infant, all is said. The infant must undergo the common lot of all syphilitic infants. In any case, it may not be confided to a nurse, and it is essential to know this from the point of view of general prophylaxis.

" In this first alternative, either the maternal nursing should be prolonged, if that be possible, or else we shall be compelled to have recourse to the special procedures that I have previously indicated to you as being able to serve for the raising of syphilitic children (recourse to a syphilitic nurse, alimentation from a goat-nurse, etc.). 2. But, if, on the contrary, after three months, or, better still, after four or five months, of observation, nothing suspicious occurs upon the infant, there are strong presumptions (I say presumptions, and nothing more) that it may have escaped the hereditary influence, that it may not be syphilitic. And here we are more at liberty in our movements ; for, in case the mother is unable to continue giving it the breast, nourishment by a nurse may be permitted—not, however, without still subjecting the nursling to a surveillance close and minute, sufficient to prevent all risk of contagion."

physical catastrophes which they may entail, the moral miseries which they may expose to publicity, the scandalous actions at law which they sometimes give rise to, the humiliation and the shame which they cause in families, etc.

In all respects, then, it is important that the physician —whenever he can do so, and he often can—should curb such contagions, by circumscribing the pox in its original home, by preventing it from carrying its dangerous pollutions elsewhere. This is for him a professional obligation in relation to society, an obligation in which he should not fail.

But I foresee an objection : "Be it so," you are, perhaps, about to say to me ; "we comprehend perfectly the importance to society, to everybody, of thus circumscribing the pox in its home, and we grant you that the means proposed by you favors this end in a measure. Nevertheless, if this means has for its undeniable result the prevention of the spread of the contagion outside of its home in a family, is it not defective, dangerous even, from other points of view? 'Thus,' you say to us, 'let the child be nursed by its mother.' But what will happen from this procedure if the mother is syphilitic and the child healthy ; or, conversely, if the mother is healthy and the child syphilitic? Will not the contagion be then transmitted from the mother to the child, or from the child to the mother? Will not this syphilitic mother infect this healthy child? Or, indeed, will not this healthy mother be infected by this syphilitic child ?"

The objection, I recognize, has, indeed, its value—at least apparently. Let us discuss it, then, with all the care which it merits, in order not to leave any reservation, any uncertainty in your minds in relation to that which precedes.

Four orders of cases are possible in the situation we are now considering. Thus :

1. Either the mother and the child have both escaped the dangers of the paternal syphilis—that is to say, remain healthy.

2. Or the mother and the child have both received the infection emanating from the father—that is, have become syphilitic.

3. Or, the mother remains healthy, while the child has undergone the contamination.

4. Or else, finally and conversely, the mother is syphilitic, while the child remains healthy.

There we have—have we not?—the four alternatives, and the four only alternatives, which can or could be presented. Besides these, there is none other to be supposed, to even be imagined theoretically.

Now, let us consider each of these in detail, and see, *à propos* of each, what may be the dangers of maternal nursing, either for the mother or for the child—a discussion which may, perhaps, seem to you somewhat long and monotonous, but which is indispensable to the clear understanding of our subject.

First hypothesis : The mother and the child have both escaped the infection.

In this case, quite evidently, there is nothing to fear for either the one or the other ; for, according to the proverb, *"qui n'a rien ne donne rien."* The nursing of the child by the mother involves no danger, then, in any respect. Let us pass it by.

Second hypothesis : The mother and child are both syphilitic.

Here, again, there is no possible danger of contagion. The mother and the child, both having syphilis, have nothing to fear from each other; for syphilis does not duplicate itself; it is not twice acquired. Let us even say that in this case the maternal nursing alone can be medically acceptable; for, at no price, for no reason, should we ever permit a syphilitic infant to be confided to a healthy nurse.

Third hypothesis: Mother healthy and child syphilitic.

It is here only that the objection we are considering seems to assume a real value. For here the possibility of a contagion arises from the simple statement even of the terms of the proposition. But we say, in the first place, that this third situation rarely presents itself in practice. We have shown that it is almost exceptional to encounter a healthy mother with a syphilitic child. Almost always, syphilis in the infant implies syphilis in the mother.

Still, however rare they may be, cases of this kind have been cited, and I have observed, at least I think I have observed, a certain number. They may, then, be taken into consideration in the present discussion. Now, the question presented in such a case is the following: Will the mother who nurses her child in these conditions be in danger of receiving syphilis from it? Theoretically, one would be induced to answer in the affirmative. Wherefore, in fact, should not this mother, who is healthy, receive the infection from her child who is syphilitic?

Practically, on the contrary, one comes to an opposite conclusion. Practically, one never sees a child, syphilitic from birth (from birth, be it well understood), infect * its

---

* On the contrary, an infant born healthy, and afterward contracting syphilis from another person (it may be a nurse, for example), is ultra-contagious

mother while nursing it. Never does one encounter a case, however plausible in theory it may be, of a mother nursing her own syphilitic child and contracting syphilis from it. Let one explain this as one chooses, it matters little to us at this moment. It is always a fact, a substantial fact, which obtrudes itself in the name of clinical observation, and which involves, in our present study, a considerable interest. Pointed out long ago by an English author — Abraham Colles * — and known to us un-

for its mother. It is thus, as has been many times observed, that the contagion of the mother is occasioned by the infant, under the following conditions : An infant is born healthy, of healthy parents ; it is temporarily confided to a syphilitic nurse and receives the syphilis from her ; returning to the maternal breast it then inoculates the mother with syphilis. Cases of this kind are found signalized everywhere. I have related a number in my *Leçons sur les nourrices et les nourrissons syphilitiques*, and I think it sufficient here to simply announce the fact without supporting it by particular citations.

* It is a curious fact that I have never witnessed nor ever heard of an instance in which a child deriving the infection of syphilis from its parents has caused an ulceration in the breast of its mother."—Abraham Colles, " Practical Observations on the Venereal Disease and on the Use of Mercury," London, 1837.

I am aware that many cases have been cited in opposition to this law, or, if the word appears a little ambitious, to the proposition of Colles. What the cases in question are worth I can not say ; for my part, I have never encountered similar ones, at least up to the present time.

It is assuredly a very surprising thing to see a healthy woman nursing her infant, covered with syphilis, remaining healthy in contact with this infant, not contracting the syphilis from it. This is indeed so extraordinary that one always questions if one is not deceived, if *this woman is indeed really exempt*, if she does not escape the contagion for the simple reason that she has already been contaminated, either before or during pregnancy. In a word, one is always tempted to believe that this woman is syphilitic, but, for some reason or another, syphilis has not been detected in her at an opportune moment—that is to say, at a moment when unequivocal manifestations would surely have been attested. Such is, at least, the interpretation which physicians, who deny the paternal heredity of syphilis, give to the proposition of Colles. For them, there can be no syphilitic infant without a syphilitic mother ; for them, the infection of the infant implies the infection of the mother. " Then," say they, " there is nothing astonishing that a syphilitic infant does not infect its mother. It can not infect her, because she is already syphilitic. One syphilitic has nothing to fear from another syphilitic."

The question, in effect, would be decided in this sense if one always verified

der the name of the *law of Colles*, this singular immunity of the mother against infection from her infant has since then impressed numbers of physicians. It is, we may say, generally accepted in our day as an undeniable fact, even confirmed by an almost unanimous assent.

syphilis in the mothers of syphilitic infants. But that is precisely what one does not *always* establish. Must one believe that it exists, even when one has no proof of it? It is this conclusion which certain of our *confrères* arrive at. Mr. Hutchinson has even built a complete theory upon this basis, to which it will not be without interest to call the attention of the reader.

According to our eminent colleague, the law of Colles can find no other possible explanation than in the infection of the mother. And yet he is the first to recognize that most commonly one can not verify the signs of infection in the mother. If, then, says he, this woman be syphilitic, she must be so in a peculiar manner, according to a certain mode which admits of her being syphilitic without apparent manifestations.

Well, continues he, that is what takes place, very probably. It is to be believed that the maternal syphilis, derived *in utero* from a syphilitic fœtus, is a syphilis of a special order—a syphilis *mitigated*, tempered, modified, susceptible of not betraying itself by any external symptom, or indeed of remaining a long time latent— even indefinitely latent. Consequently, this syphilis may escape us, may elude all our investigations, when, nevertheless, it exists, and infects the maternal organism so profoundly as to render it refractory to subsequent contamination.

As an argument for the support of this more than bold hypothesis, Mr. Hutchinson recalls to mind that virulent diseases manifest an evolution and a gravity quite different, according to their *mode of penetration* into the economy. See, says he, the variolous virus. Introduced into the organism by way of inoculation, it only determines an affection comparatively light, which results in death only once in five hundred times. Absorbed by inhalation, on the contrary, it produces a very grave disease, which becomes fatal once in four times. Apply this to syphilis, and you will easily comprehend that a syphilis derived from contamination of the fœtal blood may differ absolutely from the syphilis derived from a tegumentary inoculation, both in the symptoms of its evolution and in gravity. Developing the exposition of his theory, Mr. Hutchinson admits the possibility of three orders of cases in syphilis by conception, viz.:

"1. A first group, in which the diathesis manifests itself by the habitual symptoms of the secondary period. This is only exceptional; and it is even to be believed, according to the author, that the cases of this kind are derived from a syphilis by ordinary contagion, rather than from a syphilis by conception.

"2. A second group, in which the infection is characterized by specific symptoms, but of a light order, of a form essentially benign ; an unhealthy condition during pregnancy, loss of hair, and, later on, 'months or years later,' ulcerations of the tongue, palmar lesions, gummata of the cellular tissue.

"3. A third group (this comprehending, at least, one half of the cases), in which the disease does not betray itself by *any symptom, by any disturbance of the health.*

"It is certain," M. Ricord has written, "that, in the case
where the mother has escaped syphilis while carrying a
syphilitic infant in her womb, she never afterward contracts the syphilis by nursing her diseased infant." Likewise, M. Diday: "Never does an infant, syphilitic from
birth, communicate the disease to the mother who nurses
it." As to myself, I have never, to this day, observed a
single well-authenticated fact in contravention of the law
of Colles, and I hold this law as absolutely consistent with
the results of the clinic.

Then, to return to our subject, here again, even in this
situation, so perilous in appearance, of a healthy mother

This absence of all symptoms during the first years does not exclude the possibility of tertiary accidents in a future, more or less distant. But, most often, nothing is produced, and the syphilitic woman infected in this way generally remains
free from all specific manifestations during her life."—(" On Colles's Law and on
the Communication of Syphilis from the Fœtus to its Mother," " Medical Times
and Gazette," December, 1876, p. 643.)

I will not stop here to discuss this theory, for, to speak truly, it defies at
present all criticism. It would be necessary, in fact, either for its verification or
its refutation, to bring a whole series of clinical facts minutely observed in a
special direction, and we are not prepared with a criterion of this kind. This is a
new field of investigation which is opened to us, but in which the first landmarks
are hardly yet placed.

I shall consider it my duty, however, to mention an interesting observation
which has just been communicated to me by Dr. Charrier, and which confirms, in
one point, the doctrine of Mr. Hutchinson. The reader will find this observation
reproduced among the " Notes and Illustrative Cases " appended to this work
(Note VI).

Definitely, in the present state of science, two important facts result from
clinical observation, viz.:

1. That a healthy woman becoming *enceinte* from contact with a syphilitic man
may give birth to a syphilitic infant, while she remains healthy (in appearance, at
least).

2. That this woman, nursing a syphilitic infant, is not liable to receive the contagion from it.

It only remains to interpret this singular immunity, and, notably, to determine
if it be explicable, as certain authorities pretend, by a sort of *special* and *latent* infection of the mother—an infection derived from the fœtus by a not less special
mode of contamination. This the future alone can teach us, and we are compelled
at present to reserve judgment.

exposed to the contact of a syphilitic infant, maternal nursing does not involve any danger.

There remains, finally, a fourth and last alternative: mother syphilitic and child healthy.

This is the preceding situation reversed. Well, as in the preceding case, contagion is not exerted here. A child born healthy, although the offspring of syphilitic parents, has never taken the syphilis in nursing from its mother. As to myself, I declare that I have never seen anything of this character. I declare that I am not aware of a single example of a mother having given birth to a healthy infant, then afterward infecting it in the capacity of nurse.*

Recapitulating, then, gentlemen, whichever it may be of the four alternatives that we are considering, always and invariably we find that the nursing of the infant by its mother is free from the theoretical dangers which might be supposed.

From this an easy conclusion is to be deduced: it is that on no account should one oppose in either case the mother nursing her infant.

Now, as from the other point of view there is a valid reason, superior to every other, against the infant being confided to a nurse, the question is resolved essentially the same in two different ways. And we come to this as a final conclusion:

That under such conditions the *maternal nursing is the only rational and practical means for the raising of the infant.*

---

* It is to be well understood—and I only insist, to avoid a shade even of ambiguity—that I speak here of a mother having contracted syphilis either before or during pregnancy. For a mother contracting syphilis after her accouchement is ultra-contagious for her infant. That is a fact of common observation. It will be sufficient to recall, in this connection, those cases so numerous in which one has seen the unfortunate nurse, after having contracted syphilis from a syphilitic nursling, afterward communicate it to her own infant.

Given the case of an infant syphilitic or only suspected of syphilis, it is the mother of this infant who alone may and ought to serve as its nurse.

This is not doubtful; for a number of reasons which I can not here unfold, it does not admit of discussion. Such is the law. And, besides, I will add, in conclusion, that even if, in such circumstances, the maternal nursing should be attended with some danger, either for the mother or for the child, this consideration would in no wise modify the duty which is imposed upon the physician toward society. This duty, in any state of the case, would none the less continue to exist.

In this hypothesis, that is to say, if the maternal nursing should offer some danger, it would be your duty as physicians to contend with this new difficulty, to devise some expedient by which the possibility of contagion from the mother to the infant, or from the infant to the mother, might be averted. But we should not on that account be relieved from the strict and imperative obligation which a respect for the health of others imposes upon us. At no price, on no account, should we consent that an infant syphilitic, or even only suspected of syphilis, be confided to a healthy nurse.

The protection of society constitutes, then, in this respect—I repeat it again, and I can not too often repeat it —the capital, predominant indication, superior to every other consideration—and this, because this indication responds to interests of a general order, because it tends to a result which ought to be the aim of our common and constant efforts, viz., *to prevent the diffusion of the pox* by confining it to its sources of origin, by preventing it from being spread abroad and disseminating its germs of contagion.

NOTES AND ILLUSTRATIVE CASES.

NOTE I.

" . . . . For my part alone, I have in hand (to speak only of recorded cases) eighty-seven observations relative to syphilitic subjects, undoubtedly syphilitic, who, having married, have never communicated to their wives the least suspicious phenomenon, and, what is more, have begotten—these eighty-seven—a total of one hundred and fifty-six children, absolutely healthy " (page 19).

From its great importance, this proposition dominates the whole subject developed in this book. On this account I have judged it indispensable to legitimize it by an *exposé* of the facts from which it is deduced. I can not here relate *in extenso* these eighty-seven observations, some of which are quite long. But I will at least furnish an abstract of them, assuredly quite concise, but sufficient, I think, to fix the conviction of the reader :

Case I.—Indurated chancre of the glans ; roseola ; buccal syphilides ; epididymitic sarcocele ; treatment active and prolonged ; marriage six years after the début of the syphilis ; wife remaining uncontaminated ; three healthy children, the eldest of whom is now six years old ; gumma of the penis after the birth of the second child ; resumption of treatment.

Case II.—Indurated chancre ; roseola ; palmar syphilides ;

buccal syphilides ; treatment quite long but irregular ; marriage six years after the début of the infection ; second marriage some years later ; both wives remaining uncontaminated ; five children from these two marriages, all absolutely healthy ; recurrence of palmar psoriasis after the birth of the first and third child.

CASE III. — Indurated chancre ; roseola ; buccal syphilides ; recurrence of the roseola in circinate form ; treatment prolonged ; marriage three years after the début of the syphilis ; wife remaining uninfected ; two healthy children, the eldest of whom is now nine years of age ; ulcerated tubercle of the penis nine years after marriage.

CASE IV. — Indurated chancre ; roseola ; buccal syphilides ; eruption of crusts on the hairy scalp ; palmar and plantar syphilides ; recurrence of roseola; treatment quite prolonged ; marriage four years after début of the syphilis ; wife remaining uninfected ; a healthy child, now seventeen years old ; later, dry, tubercular syphilide (form benign).

CASE V.—Indurated chancre of prepuce ; roseola ; buccal and genital syphilides ; iritis ; average treatment ; marriage three years after début of syphilis ; wife remaining healthy ; four healthy, well-developed children.

CASE VI.—Indurated chancre ; roseola ; buccal syphilides with multiple recurrences ; treatment irregular ; marriage six years after début of the infection; wife remaining uninfected ; · five healthy children.

CASE VII.—Indurated chancre of the glando-preputial furrow; immediate ˙treatment, active and prolonged ; no other accident than roseola ; marriage after *nineteen months* of infection ; wife remaining healthy ; a healthy child now aged six years.

CASE VIII.—Indurated chancre ; papular syphilide ; recurrence five years later of an erythemato-papular syphilide; considerable treatment (pills of proto-iodide during four years, etc.); marriage eight years after début of syphilis ; wife remaining healthy ; three healthy children.

14

Case IX.—Indurated chancre of penis; papular syphilide. Crusts of hairy scalp; buccal syphilides with quite numerous relapses; prolonged methodic treatment; marriage in the third year; wife remaining uninfected; two healthy children.

Case X.—Primary accident unperceived; cutaneous and mucous syphilides; treatment from six to eight months; marriage after eight years of infection; wife remaining uninfected; five healthy children, the eldest of whom is now twelve years old; tubercular syphilide of the thorax after birth of third child; gumma of the palatine arch after the birth of fifth child.

Case XI.—Indurated chancre; papular syphilide; crusts of hairy scalp; tonsillar syphilides; treatment prolonged; marriage five years after début of infection; wife remaining healthy; two healthy children.

Case XII.—Indurated chancre; mucous syphilides; cephalalgia; methodic treatment; marriage eleven years after the début of the infection; wife remaining healthy; a healthy child, at present nine years of age.

Case XIII.—Chancre of the nose; mucous syphilides; cervical adenopathies; treatment for some months; marriage the third year; wife remaining healthy; three healthy children.

Case XIV.—Indurated chancre of the penis; roseola; buccal and anal syphilides; prolonged treatment by iodide of potassium, without mercury; marriage six years after the début of the disease; wife remaining healthy; a healthy child; début of cerebral syphilis (apoplectiform stroke, hemiplegia, etc.) five months after marriage, four months before the birth of the child.

Case XV.—Indurated chancre; roseola; buccal syphilides; short treatment; marriage nine years after infection; wife remaining healthy; a healthy child.

Case XVI.—Indurated chancre; papular syphilide; buccal syphilides with multiple recurrences; five years later, nasal caries; frightful ozæna; treatment extremely energetic during a number of years; marriage nine years after the début of the infection; wife remaining healthy; a healthy child.

CASE XVII.—Indurated chancre; roseola; mucous syphilides; eighteen months' treatment; marriage eleven years after the début of the syphilis; wife remaining healthy; two healthy children; papulo-tubercular syphilide and costal periostosis after the birth of the two children.

CASE XVIII.—Indurated chancre of the thumb; papular syphilide; tonsillar syphilides; cephalalgia; treatment energetic and prolonged; marriage four years after début of syphilis; wife remaining healthy; three healthy children.

CASE XIX.—Indurated chancre; erythemato-papular syphilide; buccal syphilides; multiple nervous accidents; anæmia; asthenia; treatment active and prolonged; marriage six years after début of syphilis; wife remaining healthy; four healthy children; after the birth of these children, cerebro-spinal accidents, very probably of specific origin.

CASE XX.—Indurated chancre; roseola; buccal syphilides; treatment for six months; marriage fifteen years after début of syphilis; wife remaining healthy; a healthy child; one month after the birth of the child accidents of cerebral syphilis; death.

CASE XXI.—Indurated chancre; papular syphilide; patches on the tonsils; treatment of about one year; marriage seven years after début of syphilis; wife remaining healthy; two healthy children (the eldest now about fifteen years old); début of cerebral syphilis three years after the birth of second child; death.

CASE XXII.—Indurated chancre; various secondary accidents; treatment of one year; marriage eight years after début of syphilis; wife remaining healthy; a healthy child.

CASE XXIII.—Indurated chancre; papulo-crusted syphilide; ecthymatous syphilide (ecthyma deep); rupia; violent cephalalgia; hemiplegia; recurrence of rupial syphilides; treatment very energetic, very prolonged; marriage two years after début of syphilis; wife remaining uninfected; healthy child; later, diplopia ; ephemeral attacks of right hemiplegia ; nasal syphilides ; ecthyma of the legs.

CASE XXIV.—Indurated chancre ; buccal syphilides ; treatment for some months ; marriage eleven years after début of syphilis ; wife remaining uninfected ; two healthy children. After the birth of these two children tibial periostosis and specific glossitis.

CASE XXV.—Indurated chancre ; roseola ; buccal syphilides ; active, prolonged treatment ; marriage fourteen months after début of syphilis ; wife remaining uninfected ; two healthy children.

CASE XXVI.—Indurated chancre ; buccal syphilides, multiple and relapsing onyxis ; papulo-squamous circinate syphilide ; periostosis ; tubercle of the glans ; treatment active and prolonged ; marriage nine years after début of syphilis ; wife remaining uninfected ; healthy child.

CASE XXVII.—Indurated chancre ; papulo-squamous syphilide ; secondary costal periostosis ; ecthyma ; tibial exostosis ; treatment prolonged ; marriage six years after début of syphilis ; wife remaining healthy ; one healthy child.

CASE XXVIII.—Indurated chancre ; slight secondary accidents ; prolonged treatment ; marriage four years after début of syphilis ; wife remaining healthy ; two healthy children.

CASE XXIX.—Indurated chancre ; roseola; buccal syphilides ; prolonged treatment ; marriage three years after début of syphilis; wife remaining healthy ; healthy child ; multiple accidents after birth of child ; specific sarcocele, periostosis, nasal ulcerations, tubercular syphilide of the nose ; diabetes.

CASE XXX.—Indurated chancre of glando-preputial furrow ; papular syphilide ; circumscribed ecthyma ; specific hydrarthrosis ; active treatment ; marriage three years after début of disease ; wife remaining uninfected ; three healthy children.

CASE XXXI.—Indurated chancre ; roseola ; crusted eruption of hairy scalp ; buccal syphilides with multiple relapses ; alopecia ;

treatment methodic and prolonged ; marriage three years after début of syphilis ; wife remaining uninfected ; a healthy child.

CASE XXXII.—Indurated chancre ; palmar psoriasis ; treatment for several months ; marriage six years after début of syphilis ; wife remaining healthy ; a healthy child ; ulcerating laryngitis, manifestly specific, three years after birth of child.

CASE XXXIII.—Indurated chancre; some secondary accidents of benign form ; later nasal osteitis, perforation of the septum ; treatment not prolonged ; marriage after five years of the disease ; wife remaining uncontaminated ; four healthy children ; fatal cerebral syphilis ; the last child was procreated after the début of the cerebral accidents (epileptiform attacks, psychical troubles).

CASE XXXIV.—Primitive accident not recognized ; roseola ; eruption of crusts of hairy scalp ; eight to ten months' treatment ; marriage twelve years after début of the disease ; wife remaining uninfected ; four healthy children ; fronto-parietal exostosis occurring a short time after the birth of the fourth child.

CASE XXXV.—Indurated chancre ; secondary angina ; cervical adenopathies ; mercurial treatment of some months ; marriage eleven years after début of syphilis ; wife remaining uninfected ; a healthy child ; syphilitic accidents of the cord preceding by one year the birth of the child.

CASE XXXVI.—Indurated chancre of the penis ; roseola ; tonsillar, lingual, palatine syphilides ; active treatment ; marriage one year after début of syphilis ; prolonged treatment after marriage ; wife remaining uninfected ; two healthy children.

CASE XXXVII.—Indurated chancre ; no secondary accidents remarked, except perhaps an anal papule ; mercurial treatment from three to four months ; marriage nine years after début of syphilis ; wife remaining uninfected ; a healthy child ; some months before the birth of the child, début of cerebral syphilis.

CASE XXXVIII.—Indurated chancre ; roseola ; cervical adenopathies ; mercurial treatment of six months ; marriage five years after début of syphilis ; wife remaining uninfected ; three healthy children ; exostoses one year after the birth of the third child.

CASE XXXIX.—Indurated chancre ; papular syphilide ; mucous patches of the tongue ; prolonged treatment ; marriage eight years after début of syphilis ; wife remaining healthy ; healthy child.

CASE XL.—Indurated chancre ; buccal mucous patches ; treatment quite protracted, principally of iodide of potassium ; marriage five years after début of syphilis ; wife remaining uninfected ; three healthy children (the eldest now aged seven years).

CASE XLI.—Indurated chancre ; roseola ; buccal syphilides ; prolonged treatment ; marriage five years after début of syphilis ; wife remaining uninfected ; two healthy children.

CASE XLII.—Indurated chancre ; cutaneous eruptions ; buccal syphilides ; ecthyma ; treatment quite prolonged ; marriage four years after début of syphilis ; wife remaining uninfected ; healthy child.

CASE XLIII.—Indurated chancre of index-finger; erythemato-papular syphilide ; alopecia ; tonsillar, labial, and lingual patches ; multiple adenopathies ; cephalalgia ; neuralgia ; prolonged treatment ; marriage four years after début of syphilis ; wife remaining uninfected ; two healthy children.

CASE XLIV.—Indurated chancre ; no other secondary accidents except buccal mucous patches ; treatment prolonged ; marriage four years after début of syphilis ; wife remaining uninfected ; two healthy children ; after the birth of the last child, cranial exostosis, with incessant relapses.

CASE XLV.—Parchment chancre of the prepuce ; papular syphilide ; specific icterus ; buccal syphilides ; prolonged treatment ; marriage eight years after début of syphilis ; wife remaining uninfected ; two healthy children.

CASE XLVI.—Indurated chancre ; buccal syphilides ; specific sarcocele ; treatment prolonged ; marriage nine years after début of syphilis ; wife remaining uninfected ; two healthy children (the eldest now aged nine years).

CASE XLVII.—Two indurated chancres of the sulcus ; erythemato-papular syphilide ; buccal syphilides ; ecthymatous syphilide of the legs ; ulcerating syphilide of the palate ; marriage in the course of the third year after début of the infection ; treatment very energetic and protracted ; wife remaining uninfected ; healthy child ; dry tubercle of the penis some months after birth of child.

CASE XLVIII.—Indurated chancre ; buccal syphilides ; alopecia ; circinate syphilides of the tongue ; treatment active and prolonged ; marriage four years after début of syphilis ; wife remaining uninfected ; healthy child ; some months after birth of child, palmar and plantar syphilides of papulo-squamous form.

CASE XLIX.—Indurated chancre ; roseola ; palmar psoriasis ; buccal syphilides ; iodide treatment ; no mercury ; marriage four years after début of the malady ; wife remaining uninfected ; two healthy children ; after birth of second child patient infects his wife from a buccal syphilide ; a pregnancy, happening the following year, terminates in an abortion.

CASE L.—Two indurated chancres ; eruptions of crusts of hairy scalp ; secondary angina ; choroiditis ; buccal syphilides ; treatment protracted ; marriage four years after début of syphilis ; wife remaining uninfected ; healthy child.

CASE LI.—Seven indurated chancres ; roseola ; impetiginous syphilide of hairy scalp ; treatment quite long ; marriage after seven years of the malady ; wife remaining uninfected ; two healthy children ; after the birth of second child, ecthymatous syphilide and gumma of palatine arch.

CASE LII.—Parchment chancre of prepuce ; roseola ; buccal syphilides ; cephalalgia ; digital psoriasis ; prolonged treatment ; marriage in the third year of the disease ; wife remaining uninfected ; healthy child.

CASE LIII.—Indurated chancre ; papular syphilide ; gummous

syphilide of pharynx; diplopia; treatment from eight to ten months; marriage ten years after début of infection; wife remaining uninfected; healthy child; one year after birth of child, début of cerebral syphilis.

CASE LIV.—Indurated chancre; circinate syphilide of hairy scalp, papulo-crusted; prolonged treatment; marriage in the fourth year of the malady; wife remaining uninfected; two healthy children.

CASE LV.—Indurated chancre; buccal syphilides; alopecia; papular syphilide; treatment of several months; marriage four years after début of disease; wife remaining healthy; healthy child; one year after birth of child, début of cerebral syphilis.

CASE LVI.—Indurated chancres; roseola; acneiform syphilide; buccal syphilides with frequent recurrences; tibial exostosis; prolonged treatment; marriage four years after début of syphilis; wife remaining uninfected; two healthy children.

CASE LVII.—Indurated chancre; buccal syphilides; superficial sclerotic glossitis; prolonged treatment; marriage in the third year of the disease; wife remaining uninfected; healthy child.

CASE LVIII.—Indurated chancre; no secondary accidents remarked; iodide treatment; gummous syphilides of the palatine arch and pharynx; frightful phagedena of the arch, the pillars, the tonsils, the pharynx; energetic treatment, prolonged several years; marriage five years after début of syphilis; wife remaining uninfected; healthy child.

CASE LIX.—Indurated chancre; roseola; buccal syphilides; cephalalgia; papular circinate syphilide; treatment prolonged; marriage six years after début of syphilis; wife remaining uninfected; healthy child.

CASE LX.—Indurated chancre; buccal syphilides; crusts of hairy scalp; treatment prolonged; marriage eight years after début of syphilis; wife remaining uninfected; healthy child.

CASE LXI.—Indurated chancre; buccal syphilides; ecthyma

of the foot; treatment of some months only; marriage three years afer début of syphilis; wife remaining uninfected; healthy child.

CASE LXII.—Indurated chancre; roseola; cephalalgia; buccal syphilides; psoriasiform syphilide; treatment prolonged; marriage four years after début of syphilis; wife remaining uninfected; healthy child.

CASE LXIII.—Indurated chancre; roseola; buccal syphilides; cephalalgia; treatment prolonged; marriage four years after début of syphilis; wife remaining uninfected; healthy child.

CASE LXIV.—Indurated chancre; roseola; buccal syphilides; treatment very protracted; marriage five years after début of syphilis; wife remaining uninfected; healthy child.

CASE LXV.—Indurated chancre; roseola; buccal syphilides; treatment of about one year; marriage in fourth year of the disease; wife remaining uninfected; healthy child; specific sarcocele at the moment of birth of child.

CASE LXVI.—Indurated chancre; cutaneous and mucous syphilides; several months' treatment; marriage eight years after début of syphilis; wife remaining uninfected; three healthy children; twelve years after marriage, paralysis of the sixth pair of nerves.

CASE LXVII.—Indurated chancre; no other secondary accidents remarked besides buccal syphilides; treatment of one year; marriage in the second year of the disease; wife remaining uninfected; three healthy children; afterward, sclerous glossitis.

CASE LXVIII.—Indurated chancre; roseola; ecthymatous syphilide of the legs; treatment quite prolonged; marriage four years after début of syphilis; wife remaining uninfected; two healthy children.

CASE LXIX.—Indurated chancre; roseola; buccal syphilides; prolonged treatment; marriage ten years after début of syphilis; wife remaining uninfected; healthy child.

CASE LXX.—Indurated chancre; papular syphilide; buccal

and anal syphilides ; iritis ; treatment prolonged ; marriage five years after début of syphilis ; wife remaining uninfected ; two healthy children.

CASE LXXI.—Indurated chancre ; buccal syphilides ; eruption on hairy scalp ; treatment quite prolonged ; marriage eight years after début of syphilis; wife remaining uninfected ; a healthy child.

CASE LXXII.—Indurated chancre ; buccal syphilides ; palmar psoriasis ; four months' treatment ; marriage five years after début of syphilis ; wife remaining uninfected ; healthy child ; afterward, gumma of the palatine arch.

CASE LXXIII.—Indurated chancre ; diverse secondary accidents ; treatment of some months ; six years later, palmar psoriasis ; resumption of treatment ; marriage thirteen years after début of syphilis ; wife remaining uninfected ; two healthy children.

CASE LXXIV.—Indurated chancre ; buccal syphilides ; palmar psoriasis ; energetic treatment ; marriage in the course of the second year of the disease ; wife remaining uninfected ; healthy child.

CASE LXXV.—Indurated chancre ; cutaneous and mucous syphilides ; treatment irregular, still sufficiently prolonged ; marriage four years after début of syphilis ; wife remaining uninfected ; a healthy child ; two years later, papulo-crusted syphilide of circinate form.

CASE LXXVI.—Indurated chancre ; cutaneous syphilides ; treatment of some months ; marriage five years after début of syphilis ; wife remaining uninfected ; two healthy children ; four years after birth of second child, début of cerebral syphilis.

CASE LXXVII.—Indurated chancre ; roseola ; buccal syphilides ; ecthyma of the leg ; treatment prolonged ; marriage two years after début of syphilis ; wife remaining uninfected ; a healthy child.

CASE LXXVIII.—Indurated chancre ; no secondary accidents

remarked ; mercurial treatment for six months ; marriage three years after début of syphilis ; wife remaining uninfected ; three healthy children ; after the birth of the third child, début of locomotor ataxia.

CASE LXXIX.—Labial chancre ; cutaneous syphilides ; no treatment ; marriage seven years after début of syphilis ; wife remaining healthy ; healthy twin children ; after the birth of these two children, palatine gumma, tertiary ulcerations of the nasal fossæ, cephalalgia.

CASE LXXX.—Indurated chancre ; cutaneous and mucous syphilides ; palmar psoriasis ; prolonged treatment ; marriage four years after début of syphilis ; wife remaining uninfected ; a healthy child.

CASE LXXXI.—Indurated chancre ; papular syphilide ; buccal and genital syphilides ; treatment prolonged ; marriage six years after début of syphilis ; wife remaining uninfected ; two healthy children.

CASE LXXXII.—Indurated chancre of glans ; marriage almost immediately after cicatrization of chancre ; various secondary accidents, buccal syphilides, palmar psoriasis ; treatment prolonged ; the patient avoiding any fecundating connection during five years ; the following year a healthy child ; wife remaining uninfected ; consecutive to the birth of the child, tibial periostosis ; cerebral syphilis.

CASE LXXXIII.—Primary accident not recognized ; roseola in 1866 ; buccal syphilides ; prolonged treatment ; marriage six years after début of syphilis ; wife remaining uninfected ; healthy child.

CASE LXXXIV.—Indurated chancre ; buccal syphilides ; prolonged treatment ; marriage fourteen years after début of disease ; wife remaining uninfected ; a healthy child.

CASE LXXXV.—Indurated chancre ; roseola ; palmar psoriasis ; treatment prolonged ; marriage seven years after début of syphilis ; wife remaining uninfected ; a healthy child.

CASE LXXXVI.—Indurated chancre; various secondary accidents; treatment for fourteen weeks, composed principally of iodide of potassium; but little mercury; marriage five years after début of syphilis; wife remaining uninfected; a healthy child, at present fifteen years old; fifteen years after marriage, tuberculo-ulcerative syphilide of nose.

CASE LXXXVII.—Indurated chancre; papular syphilide; buccal syphilides; genital syphilides; treatment prolonged; marriage nine years after début of disease; wife remaining uninfected; a healthy child.

Independently of the principal demonstration furnished by the preceding statistics, they bring into prominence a most important fact, viz., that syphilitic subjects may be inoffensive in marriage to their wives and children, even while they remain under the power of the diathesis, and are destined to undergo new attacks. And, in effect, these statistics embrace no less than *thirty-five* cases of this order, in which various accidents of a nature incontestably specific occurred after marriage, without, however, the wives and the children of these different patients having suffered the least bad result medically. This fact is certainly reassuring. Nevertheless, one should not exaggerate the importance of these statistics, nor attach to them a signification of which they do not admit. While it is true that the patients in question have transmitted, hereditarily, nothing to their offspring, and have communicated nothing to their wives (which is explained by the situation or the character of their accidents), they have no less been, a certain number at least, very prejudicial to their families on account of the *personal* consequences of their disease. Many, for example, have died; others have only survived with functional troubles, more or less important, with serious infirmities, etc., and that to the great detriment of the social community constituted by marriage.

On the other hand, note it well, the preceding statistics have neither for object nor for result, from the point of view of the protection of wives and children, the establishment of numerical relations between the subjects who marry after a sufficient depuration and those who contract marriage in conditions precisely opposite. This relation necessarily .escapes us, and will always escape us. In effect, we only take cognizance of those patients who come to us on account of various accidents, and these are always certain to find a place in our statistics ; while the others remain unrecognized by us, for the excellent reason that, having no further manifestations of syphilis, they have no occcasion to claim our services.

Finally, the preceding statistics show us certain examples of syphilis particularly grave, which have, nevertheless, remained inoffensive in marriage, at least so far as relates to the dangers incurred by the wife and the children. Case XXIII is a type of this class. This, assuredly, was one of the cases in which every prudent physician would have considered it his duty to *interdict marriage*, by reason of the multiplicity and the threatening character of the manifestations (profound ecthymatous syphilide with multiple recurrences, rupia, the most violent cephalalgia, hemiplegia, etc.). The result, however, has not justified the apprehensions which the gravity of the symptoms was calculated to excite.

NOTE II.

SYPHILIS.—SEVEN ABORTIONS OR PREMATURE ACCOUCHE-
MENTS.

X——, aged forty years, seamstress, entered the Lourcine Hospital, June 16, 1870.

She is a woman of tall figure, who appears to have
formerly had a robust constitution, but who has become,
according to her statement, very much enfeebled by work,
grief, and numerous pregnancies.

She has always enjoyed excellent health. She even
boasts that she has never suffered, apart from her confinements, the least indisposition.

Married at nineteen years of age, she, first of all, had
three "superb children," two of which are still living and
in excellent health. The third, which was likewise well
developed, died in infancy, and appears to have succumbed to some incidental malady of an acute form
(probably pneumonia). At the age of twenty-nine years,
this woman contracted syphilis from her husband, who
had himself contracted it quite recently. At the same
time she became *enceinte*. This pregnancy was terminated by an abortion in the fifth month. As accidents
of syphilis, the patient states that she had, at first,
an indurated chancre of the vulva; then, soon afterward,
an eruption of small red spots, which rapidly covered the
body, the limbs, and the lower portion of the face.

Later she had new papules on the skin, erosions in the mouth, and, especially, a very tenacious eruption on the palms of the hands. This eruption had been called psoriasis by a physician. It continued not less than a year.

On account of these various accidents the patient twice entered the Hospital St. Louis, in the service of Dr. Gibert. The second time she remained nearly six months. She remembers having been treated by a mercurial syrup, then by a solution of iodide of potassium. Since then she has not had any treatment, although at times she has experienced new accidents, notably ulcerations in the mouth, violent pains in the arms and in the back, diffused neuralgias, and a very characteristic sciatica.

She has not been better treated, she says, because she has scarcely ceased being *enceinte* since that time. And, in reality, from that date until 1867, she has had no fewer than six pregnancies, which all terminated disastrously, as follows :

Fifth pregnancy : premature accouchement at seven months and a half; child sickly, stunted, dying on the fifteenth day.

Sixth pregnancy : accouchement almost at term ; child still-born.

Seventh pregnancy : premature accouchement at seven months and a half ; child still-born. The patient's mother, who assisted at the confinement, said to her that the child's skin "was quite black and came off in pieces."

Eighth pregnancy : premature accouchement; child still-born.

Ninth pregnancy : abortion at three months and a half.

Tenth pregnancy : abortion at six weeks, accompanied with considerable hæmorrhage, and followed by several metrorrhagias.

To recapitulate, then : *ten* pregnancies, of which the *three* anterior to the syphilis resulted, at term, in healthy children, and the *seven* posterior to the syphilis terminated in four premature accouchements and three abortions.

Within the last two years new accidents again appeared, viz. : "a tumor" at the level of the left clavicle, quite voluminous and very sensitive to pressure ; an eruption of crusts on the hairy scalp ; an abundant loss of hair, etc. These various accidents brought the patient to Lourcine, where she was treated (service of Dr. Péan) with mercurial pills and the iodide of potassium. She left cured ; even her hair had almost entirely grown in again.

Outside, the patient continued the medication for several months, returning from time to time to the hospital for consultation, where we saw her for the first time.

Finally, about a month ago, she felt two lumps, "like two kernels," which were formed in the tongue. A third kernel soon formed in the neighborhood of the two others. Then all three became ulcerated, and on this account she came to ask our care.

To-day, we find upon the extremity of the tongue three well-circumscribed ulcerations, with borders adherent and clearly cut, with grayish bottom, the base engorged and renitent. In aspect, they are types of *gummous lesions*. No symptomatic adenopathy. No other accidents.

Treatment : iodide of potassium, in a daily dose of from three to five grammes, progressively increased ; painting twice a day with the tincture of iodine ; gargles of infusion of marsh-mallow, and pulverizations of iodide solution upon the tongue. Rapid cure.

NOTE III.

"It may be said very positively, and without any exaggeration, that the syphilitic influence of the mother is veritably *pernicious* for the fœtus." (Page 63.)

The following statistics, collected from different sources, and which, for reasons above mentioned, I intentionally give separately, go to establish this with a numerical evidence unfortunately too complete :

I.

, The first relates to syphilitic women observed in this city, in private practice. It comprises eighty-five cases of pregnancy, which, considered only in their result, the most direct and the least subject to error, viz., the *death* or the *survival* of the child, have furnished me with the following figures :

Cases of survival.................................... 27
Cases of death (abortions, premature accouchements, still-born
    infants, infants dead within a short time after delivery).. 58
                                                                ——
        Total.................................... 85

Here are the details of these different cases :

Case I.—X——, nineteen years old ; indurated chancre of the
    15

lip, misunderstood as to its nature ; papulo-squamous syphilide ; no treatment ; miscarriage at third month.*

Case II.—Twenty-one years ; contagion at beginning of marriage ; roseola ; papulo-erosive syphilides of the vulva and anus ; mercurial treatment for some months ; pregnancy ; accouchement at about seven months ; child very miserable, dying on fifth day.

Case III.—Twenty-five years ; début of syphilis unknown ; pregnancy ; mercurial treatment of short duration ; accouchement at term ; child syphilitic ; energetically treated and surviving (now nine years old).

Case IV.—Twenty-eight years ; début of syphilis unknown ; various secondary accidents ; prolonged treatment (mercury and iodide of potassium); pregnancy six years after first accidents ; accouchement at term ; child healthy and surviving ; two years after birth of child, superficial syphilides of the tongue.

Case V.—Thirty-one years ; contagion at date of marriage ; treatment very irregular and of short duration ; four pregnancies in five years ; first pregnancy : accouchement at seven and a half months, child very small, cachectic, born with a specific eruption, and dying in some hours ; the three other pregnancies were terminated by abortion.

Case VI.—Twenty-five years ; husband syphilitic ; pregnancy from beginning of marriage ; syphilis by conception ; papulo-squamous syphilide ; vulvar and buccal syphilides ; mercurial treatment for some weeks ; accouchement before term ; child born with a syphilitic eruption ; dying on eleventh day.

Case VII.—Twenty-one years ; infected soon after marriage (syphilis by conception at least probable); specific treatment long time continued ; first child syphilitic, surviving ; second pregnancy

---

* I shall only mention here, be it well understood, miscarriages absolutely *spontaneous*—I mean happening without an accidental cause, and which could not reasonably be imputed to specific influence. I have rigorously excluded from these statistics all cases where there existed the least suspicion of the possible action of any cause whatever outside of syphilis.

terminates by abortion (accidental causes alleged); third and fourth pregnancies terminating at term ; children healthy and living.

CASE VIII.—(See page 117.)

CASE IX.—Infected by her husband in the last months of first pregnancy ; vulvar chancre ; roseola ; cephalalgia ; specific treatment quite prolonged ; accouchement at term ; child syphilitic, dying in a few hours ; second pregnancy : accouchement at eight months ; child presenting syphilitic spots at birth ; dying in half an hour ; third pregnancy : accouchement at term ; child healthy in appearance, dying suddenly of convulsions at seven months ; fourth pregnancy : accouchement at term ; child healthy ; surviving; fifth pregnancy : accouchement before term ; child dying in a few hours ; sixth pregnancy : abortion ; seventh pregnancy : accouchement at term ; child healthy ; surviving.

CASE X.—Twenty-three years ; date of origin of syphilis unknown ; various secondary accidents ; treatment of some weeks ; accouchement at term ; child syphilitic, infecting its nurse, and dying at the age of one month ; second pregnancy five years later, after prolonged treatment ; child healthy ; surviving.

CASE XI.—Infected by her husband ; various secondary accidents ; treatment of short duration ; three pregnancies ; first child dying at six weeks ; second child dying in three hours ; third child still-born ; at this time specific treatment, which is prolonged several years ; fourth pregnancy : child healthy ; surviving.

CASE XII.—Twenty-one years ; syphilis dating back several months ; papulo-squamous syphilide ; lingual syphilides ; onyxis ; treatment two or three months ; abortion.

CASE XIII.—Twenty-eight years ; infected by her husband ; treatment of fifteen days ; first pregnancy giving a syphilitic child, which dies in a few hours ; the husband and wife then submit themselves to a specific treatment, which is pursued during two years ; second pregnancy three years later ; child healthy ; surviving.

CASE XIV.—Seventeen years ; vulvar chancre ; papular syphilide ; palmar psoriasis ; cephalalgia ; treatment short, in very feeble doses ; pregnancy in the second year of the disease ; accouchement before term ; child dying in three weeks, in a state of frightful consumption.

CASE XV.—Twenty-seven years ; syphilis in 1869 ; confluent syphilides of vulva ; buccal syphilides ; treatment of a few weeks ; premature accouchement in 1870 of dead child ; premature accouchement in 1871 ; child dead.

CASE XVI.—(See page 118.)

CASE XVII. — Twenty-two years ; syphilis by conception ; erythemato-papular syphilide ; tonsillar syphilides ; cephalalgia ; treatment of some months ; accouchement at seven months ; child healthy in appearance ; dies suddenly after some days.

CASE XVIII. — Twenty-two years ; syphilis by conception ; secondary accidents toward the fifth month of pregnancy ; mercurial treatment ; accouchement at seven and a half months ; child affected with grave syphilis ; energetically treated, it survives.

CASE XIX.—Thirty years ; syphilis in 1872 ; chancre not recognized ; vulvar and buccal syphilides ; alopecia ; treatment of some months ; abortion in 1875.

CASE XX.—Twenty-five years ; infected from the data of marriage ; vulvar chancres ; syphilides ; four months of treatment ; two miscarriages in the first two years succeeding marriage ; the fifth year, child syphilitic ; surviving ; contamination of the nurse.

CASE XXI.—Twenty-nine years ; cutaneous syphilides ; vulvar syphilides ; treatment of some months ; pregnancy in the first months of the disease ; abortion.

CASE XXII.—Thirty years ; début of syphilis unrecognized, manifesting itself in the course of pregnancy ; papular syphilide ; palmar psoriasis ; onyxis ; treatment of few weeks ; accouchement at term ; child dead fifteenth day.

CASE XXIII.—Twenty-nine years ; début of syphilis unknown ;

no treatment; papular syphilide; buccal syphilides; abortion; afterward, periostosis and accidents of cerebral syphilis.

CASE XXIV.—Twenty-two years; accidents of secondary syphilis in the course of pregnancy; roseola; cephalalgia; intense neuralgias; retino-choroiditis; treatment of some months; abortion.

CASE XXV.—Twenty-five years; indurated vulvar chancre; roseola; lingual syphilides; treatment of some weeks; pregnancy six months after début of disease; abortion.

CASE XXVI.—Twenty-three years; infected from the début of her marriage and became *enceinte* simultaneously; treatment of some months; accouchement of a dead child at eight months; three pregnancies the three following years; treatment of some months in the course of each pregnancy; second child syphilitic; dead at two months; third child syphilitic; treated energetically, it survives; fourth child healthy; entirely well.

CASE XXVII.—Twenty-two years; indurated chancre of the buttock in 1870; papular syphilide; vulvar and buccal syphilides; frontal periostitis; ten months of regular treatment (mercury and iodide of potassium); accouchement at term, in December, 1872; child healthy.

CASE XXVIII.—Twenty years; contagion in sixth month of pregnancy; indurated vulvar chancre; tonsillar syphilides; treatment only commenced at eighth month; five days after, accouchement of a macerated fœtus.

CASE XXIX.—Nineteen years; secondary accidents appearing in the first months of pregnancy; erythemato-papular syphilide; vulvar syphilides; cephalalgia; treatment of ten months; accouchement of a dead child.

CASE XXX.—Twenty years; secondary accidents appearing in third month of pregnancy; treatment of some months; abortion; afterward, mercurial and iodide treatment long time pursued; second pregnancy two years later; accouchement at term; healthy child; onyxis consecutively to accouchement.

Case XXXI.—Twenty-seven years ; secondary accidents appearing in the course of pregnancy ; treatment of some months ; abortion ; second pregnancy : premature accouchement ; child dead on fifteenth day ; third pregnancy : accouchement at term ; child syphilitic ; treated ; surviving.

Case XXXII.—Twenty-five years ; début of syphilis unknown ; not treated ; pregnancy three to four months ; papulo-squamous syphilide ; ulceration of the tonsils ; alopecia ; abortion.

Case XXXIII.—Twenty-seven years ; début of syphilis unknown ; no treatment ; pregnancy from fourth month ; tonsillar syphilides ; osteocopic pains ; abortion.

Case XXXIV.—Twenty-three years ; secondary accidents appearing in the course of pregnancy ; no treatment ; abortion.

Case XXXV.—Twenty-six years ; syphilis transmitted by catheterism of the Eustachian tube, and remaining for a long time unrecognized ; herpetiform syphilide ; ecthyma, cephalalgia, neuralgias ; treatment of some months ; abortion.

Case XXXVI.—Twenty-five years ; syphilis of unknown début and not treated ; various secondary accidents ; first pregnancy ; child still-born ; second pregnancy : child dying fifteenth day ; consecutively, tubercular syphilide.

Case XXXVII.—Twenty-four years ; roseola ; papular syphilide ; buccal syphilides ; treatment from five to six months ; pregnancy in second year of disease ; abortion.

Case XXXVIII.—Thirty-one years ; syphilis not recognized ; cicatrices clearly specific ; no treatment ; abortion ; some years later, syphilis of the brain ; death.

Case XXXIX.—Twenty-seven years ; infected at beginning of marriage ; treatment of few months ; two pregnancies, terminating by abortion ; afterward, gumma of palatine arch and perforation of palate.

Case XL.—Twenty-five years ; contagion in third month of pregnancy ; indurated chancre of labia minora ; papular syphilide ; treatment of one month ; accouchement at term ; child syphilitic; treated; dying at eight months; treatment regular and prolonged after accouchement; three years later, second pregnancy; child healthy, surviving.

CASE XLI.—Twenty-three years; début of syphilis unknown; papulo-crusted syphilide; buccal syphilides; irregular treatment; pregnancy five years after infection; accouchement almost at term; child syphilitic, soon dying; two miscarriages the two following years.

CASE XLII.—Twenty-two years; infected in second month of pregnancy; chancre *parcheminé* of the vulva; vulvar syphilides; mercurial treatment until the end of pregnancy; accouchement at term; child surviving, never having presented but a slight eruption, nature of which remains doubtful.

CASE XLIII.—Twenty-three years; syphilitic infection and pregnancy from début of marriage; no treatment; abortion at second month.

CASE XLIV.—Twenty-two years; pregnancy from début of marriage; accidents of secondary syphilis appearing from the third or fourth month of gestation; treatment of some months; accouchement at term; child syphilitic; treated; surviving. (This child infected its nurse, who has transmitted the syphilis— 1. To her child; 2. To her husband.)

CASE XLV.—Twenty-three years; secondary accidents making invasion in fifth month of pregnancy; mercurial treatment; accouchement at term; child syphilitic; treated; surviving; contagion transmitted to the nurse.

CASE XLVI.—Twenty years; infected from début of marriage; treatment for some weeks; two abortions in the first year; recommencement of treatment, which is continued two and a half years; pregnancy four years later; accouchement at term; child healthy (at present five years old).

II.

Our second statistics have been recorded of patients observed in hospital practice, for the most part at the Lourcine, some at the Saint Louis. It has furnished us the following results:

Cases of survival of infant............................. 22
Cases of death of infant (abortions, premature accouchements, still-born, children dead shortly after the accouchement).. 145
Total... 167

Here are the cases which have furnished the elements of these statistics:

CASE I.—Twenty-seven years; syphilis of unknown origin; roseola and vulvar syphilides in 1872; no treatment; in 1875, accouchement at seven months; child still-born; in 1879, tubercular syphilide, taking on a phagedenic form.

CASE II.—Thirty years; syphilis in 1868; cutaneous eruptions; syphilides of the mucous membranes; alopecia; very short treatment; in 1869, accouchement at term of a dead child; in 1875, miscarriage at three months; in 1878, enormous gumma of the sternal region.

CASE III.—Twenty-five years; syphilis of unknown origin, but of certainly recent date; papulo-hypertrophic syphilide of the vulva; crusts of the hairy scalp; very irregular treatment; abortion at fourth month.

CASE IV.—Thirty-five years; in 1877, syphilis occurring in the course of pregnancy; papulo-hypertrophic syphilides of the vulva and the perinæum; treatment for some months; accouchement at term; child dying of convulsions when one month old; second pregnancy in 1878; abortion.

CASE V.—Twenty-one years; pregnancy eight months advanced; genital and peri-anal syphilides; no treatment until entrance to hospital; mercurial treatment; accouchement at term; child syphilitic; dying in five months.

CASE VI.—Twenty-seven years; syphilis probably hereditary; tuberculo-ulcerative syphilide of phagedenic form, having commenced when she was eight years old, and still persisting nineteen years later. This lesion has traversed the entire extent of the inferior limb, affecting a serpentine course; cured quite rapidly by specific treatment. Five pregnancies. First pregnancy, child dying at two and a half years; second pregnancy, child dying at six months; third pregnancy, child affected "with an immense sore, which involved the entire chest"—dying at three years; fourth pregnancy, child hydrocephalic, dying at five months; fifth pregnancy, child dying suddenly, and "without disease," at the age of three months.

CASE VII. — Thirty-nine years; syphilis unrecognized; in 1877, gummous syphilide destroying the nasal septum and its base, multiple and very large cicatrices disseminated over the whole body. The lesions which produced these cicatrices date back fifteen years. Two pregnancies in 1867 and 1868. First child dying at one year (cause unknown); second child dying at three weeks, in a state of consumption.

CASE VIII.—Syphilis at twenty-five years; roseola; mucous patches, alopecia, osteocopic pains, febrile attacks; later, ulcerating syphilides leaving deep cicatrices; treatment irregular. First pregnancy at thirty years; child syphilitic, surviving; second pregnancy the following year, child surviving; five years later, rupia.

CASE IX.—Twenty-three years; in 1874, pregnancy, in the course of which appeared mucous and cutaneous syphilides; accouchement at term; child dying at three months (cause unknown); second pregnancy in 1875; abortion at three months; third pregnancy in 1876; abortion at seven months; treatment always irregular. Later, in 1878, ulcerating syphilide of the vulva.

CASE X.—Twenty-five years; syphilis unknown; gumma of •

pharynx in 1878 ; the same year, accouchement before term, child dying of convulsions on eighth day.

CASE XI. — Twenty-nine years ; syphilis unknown ; buccal and vulvar syphilides ; alcoholism ; pregnancy ; child submitted to an energetic treatment, surviving.

CASE XII.—Twenty-two years ; accidents of secondary syphilis in the course of pregnancy ; treatment very short ; premature accouchement ; child still-born. The following year, second pregnancy ; accouchement at term, child dying in six weeks. The following year, third pregnancy ; accouchement at term, child dying first day. Later, papulo-squamous syphilide of circinate form.

CASE XIII.—Twenty-one years ; syphilis in 1875 ; contagion in the course of pregnancy ; indurated chancre of the vulva ; secondary angina ; neuralgias ; alopecia ; no treatment ; abortion ; second pregnancy ; premature accouchement of a still-born child. In 1878, tibial periostosis.

CASE XIV.—Twenty years ; accidents of secondary syphilis appearing in course of pregnancy ; mercurial treatment of two and a half months ; premature accouchement ; child syphilitic ; dying on tenth day.

CASE XV.—Twenty-two years ; pregnancy ; contagion at début of pregnancy ; chancre of neck of uterus ; roseola ; vulvar syphilides ; crusts of the hairy scalp.; mercurial treatment not prolonged ; accouchement before term ; child still-born.

CASE XVI.—Eighteen years ; contagion at début of pregnancy ; chancre of neck of uterus ; vulvar syphilides ; roseola ; palmar psoriasis ; treatment of a few weeks ; abortion at third month.

CASE XVII.—Twenty-two years ; pregnancy at five and a half months ; indurated chancre of the vulva ; roseola ; treatment of a few weeks ; premature accouchement ; child still-born.

CASE XVIII.—Nineteen years ; syphilis in 1869 ; chancre of the uterine neck ; cephalalgia ; cranial periostoses ; no regular treatment ; pregnancy in 1871; abortion.

Case XIX.—Twenty years; accidents of secondary syphilis in 1872; papulo-erosive syphilides of the vulva; treatment of few weeks; in 1875, pregnancy; accouchement at term; child probably syphilitic; dying in four months.

Case XX.—Twenty-seven years; début of syphilis unknown; vulvar syphilides; abortion three weeks before entrance into hospital.

Case XXI.—Seven abortions or premature accouchements (vide case in detail, p. 210).

Case XXII.—Twenty years; pregnancy, in course of which appeared various secondary accidents (roseola, vulvar syphilides, eruptions of crusts of hairy scalp, alopecia); treatment of a few weeks; abortion at six and a half months.

Case XXIII.—Twenty years, at the time of entrance into hospital; pregnancy at eighth month; papulo-hypertrophic vulvar syphilides; roseola; emaciation; alopecia; asthenia; treatment by mercurial frictions and iodide of potassium; accouchement at term; child syphilitic; dying at six weeks, in state of cachexia.

Case XXIV.—Eighteen years; contagion contemporary with début of pregnancy; papular syphilide; vulvar syphilide; treatment of six weeks; abortion at six and a half months.

Case XXV. — Twenty - two years; pregnancy at seventh month; début of syphilis unknown; confluent vulvar syphilides; alopecia; febrile attacks; mercurial treatment; accouchement at term; child dying of convulsions at five weeks.

Case XXVI.—Twenty-six years; secondary syphilis of recent origin; no treatment; abortion at fourth month.

Case XXVII.—Seventeen years; pregnancy of seven months; papulo-ulcerative syphilides of the vulva; mercurial treatment of four weeks; accouchement at eighth month; child still-born.

Case XXVIII.—Twenty-six years; pregnancy of six months; syphilis appearing to date from four months; vulvar, peri-vulvar, anal, genito-crural, and buccal syphilides; circinate roseola; treatment of few weeks ; accouchement at eighth month ; child still-born.

CASE XXIX.—Twenty-two years ; pregnancy of five and a half months ; début of syphilis unknown ; papulo-squamous syphilide ; cephalalgia ; rheumatoid pains ; specific fever ; frontal periostosis ; enteritis ; emaciation ; imminence of cachexia ; tonic and specific treatment ; accouchement at eight months, child living only two hours.

CASE XXX.—Eighteen years ; pregnancy of seven months ; indurated chancre of upper lip ; tonsillar syphilides ; mercurial treatment ; accouchement at term ; child living, healthy.

CASE XXXI.—Twenty-one years ; indurated chancre of vulva at fourth month of pregnancy ; mercurial treatment ; abortion at six months.

CASE XXXII.—Eighteen years ; pregnancy of two months ; roseola ; crusts of hairy scalp ; cephalalgia ; mercurial treatment ; abortion at three months.

CASE XXXIII.—Twenty-four years ; syphilis dating from five years ; treatment quite short ; ulcerating syphilides of the vulva ; pregnancy ; accouchement at term, child dying second day.

CASE XXXIV.—Twenty-three years ; syphilis in 1869 ; treatment of two to three months ; in 1871, pregnancy ; simple chancres and suppurating buboes ; papulo-squamous syphilides ; vulvar syphilides ; accouchement at term ; child syphilitic, dying on the fifteenth day.

CASE XXXV.—Eighteen years ; syphilitic chancre in May, 1868 ; roseola ; no treatment ; accouchement at term, in June, 1868, child living only five hours.

CASE XXXVI.—Twenty-four years ; pregnancy ninth month ; indurated vulvar chancres ; cephalalgia ; alopecia ; accouchement at term, two days after entrance into hospital ; child syphilitic ; dying at fourth month.

CASE XXXVII.—Twenty-two years ; pregnancy ; contagion

during eighth month ; syphilitic chancre of the vulva ; mercurial treatment ; accouchement almost at term ; child small but surviving (lost sight of when six weeks old).

CASE XXXVIII.—Twenty-nine years ; secondary accidents in the course of pregnancy ; treatment irregular, brief ; accouchement at eight months, child dying in six days.

CASE XXXIX.—Seventeen years ; début of secondary syphilis unknown ; no treatment ; abortion at fourth month.

CASE XL. — Twenty-one years ; pregnancy, two to three months ; papulo-erosive vulvar syphilides ; two months' treatment ; accouchement at term ; child syphilitic ; dying at fourth week.

CASE XLI.—Thirty years ; syphilitic chancre at fifth month of pregnancy ; roseola ; alopecia ; buccal syphilides ; two months' treatment ; accouchement at seven and a half months ; child stillborn.

CASE XLII.—Twenty-five years ; pregnancy, fifth month ; syphilis of unknown origin ; no treatment ; roseola ; vulvar syphilides ; abortion the second day after entrance into hospital.

CASE XLIII. — Twenty-two years ; pregnancy of eight months ; début of syphilis unknown ; genital syphilides ; cutaneous syphilides ; treatment of three weeks ; accouchement at term, child dying in five days.

CASE XLIV.—Twenty-two years ; accidents of secondary syphilis manifesting themselves in the course of pregnancy ; vulvar syphilides ; no treatment ; abortion at five and a half months.

CASE XLV.—Twenty-two years ; pregnancy of eight months ; roseola ; vulvar syphilides ; mercurial treatment ; accouchement at term ; child small, emaciated, dying at fifth day.

CASE XLVI.—Thirty years ; pregnancy of eight months ; secondary accidents manifesting themselves in the second half of pregnancy ; vulvar syphilides ; temporal periostosis ; cephalalgia ; asthenia ; emaciation ; premature accouchement ; child dying at fourth day.

CASE XLVII.—Twenty-eight years ; pregnancy of eight

months ; secondary accidents happening during pregnancy ; papular syphilides ; cephalalgia ; neuralgias ; analgesia ; no serious treatment ; accouchement at term ; child dying day of its birth.

CASE XLVIII.—Twenty-six years ; accidents of secondary syphilis occurring in second half of pregnancy ; vulvar syphilides ; treatment of a few weeks ; accouchement at seven and a half months ; child miserable, dying in a few hours.

CASE XLIX.—Twenty-seven years ; début of secondary syphilis unknown ; cutaneous syphilides ; vulvar and buccal syphilides ; no serious treatment ; accouchement almost at term ; child syphilitic, dying fifteenth day.

CASE L.—Twenty-three years ; début of syphilis unknown ; vulvar syphilides ; treatment of few weeks ; abortion at six and a half months.

CASE LI.—Nineteen years ; accidents of secondary syphilis appearing in the earlier months of pregnancy ; mercurial treatment and iodide in small doses ; accouchement at term ; child probably syphilitic, dying at one month.

CASE LII.—Nineteen years ; indurated chancre of vulva in fourth month of pregnancy ; vulvar syphilides ; specific treatment ; accouchement at term ; child dying in four days.

CASE LIII.—Twenty years ; début of syphilis unknown ; multiple secondary accidents ; no treatment ; accouchement at seven months ; child dying at two months (probably syphilitic).

CASE LIV.—Twenty-two years ; syphilis in 1869, toward the end of first pregnancy ; treatment of fifteen days ; child healthy, surviving ; in 1872 second pregnancy ; abortion at six months ; later, gummous syphilide.

CASE LV.—Twenty-five years ; début of syphilis unknown ; secondary accidents ; pregnancy ; hydramnios ; accouchement at seven months ; child still-born.

CASE LVI.—Twenty years ; accidents of secondary syphilis coincident with début of pregnancy ; vulvar syphilides ; crusts of hairy scalp; treatment of two months; accouchement at term; child small, miserable, dying in twenty-four hours.

CASE LVII.—Twenty-one years; début of secondary syphilis unknown; pregnancy of seven months; treatment of few weeks; accouchement of dead child.

CASE LVIII.—Nineteen years; pregnancy of five months; secondary syphilis appearing to date from two to three months; treatment of a few weeks; abortion.

CASE LIX.—Twenty years; pregnancy of seven and a half months; début of syphilis unknown; vulvar and tonsillar syphilides; cephalalgia; neuralgiform pains; pigmentary syphilide of the neck; treatment of a few weeks; accouchement at term; child dying in few hours.

CASE LX.—Twenty-four years; pregnancy of eight months; début of syphilis unknown; vulvar syphilides; no treatment; child still-born.

CASE LXI.—Twenty-four years; pregnancy of eight months; syphilis appearing to date back three months; vulvar syphilides; febrile attacks; treatment of one month; accouchement at term; child syphilitic, treated, surviving.

CASE LXII.—Twenty-three years; pregnancy in fourth month of syphilis; vulvar syphilides; cephalalgia; no treatment; premature accouchement; child still-born.

CASE LXIII.—Twenty-two years; pregnancy of six months; contagion during pregnancy; roseola; confluent syphilides of the vulva, of the perinæum, of the anus, of the genito-crural fold; emaciation; no treatment; abortion.

CASE LXIV.—Twenty years; pregnancy of from four to five months; début of syphilis unknown; confluent syphilides of the vulva, of the anus, of the mouth; chloro-anæmia; cephalalgia; analgesia; treatment very short and irregular; accouchement at seventh month; child dying in twelve days.

CASE LXV.—Twenty years; pregnancy of six months; secondary accidents appearing during pregnancy; confluent syphi-

lides of the vulva ; palmar psoriasis ; papulo-squamous syphilide ;
active mercurial treatment (from five to twenty grammes of the
proto-iodide daily during three months) ; accouchement at term ;
child living, apparently healthy (lost sight of when twelve days
old).

CASE LXVI.—Twenty years ; secondary syphilis of recent
date ; syphilides of the vulva and of the throat ; pregnancy ; no
treatment ; accouchement almost at term ; child syphilitic, dying
of convulsions when three weeks old.

CASE LXVII.—Forty-four years ; syphilis coinciding with
début of pregnancy ; papulo-hypertrophic syphilides of the vulva
and of the anus ; buccal syphilides ; alopecia ; papulo-squamous
syphilide, herpetiform at several points ; no treatment ; abortion
at six months.

CASE LXVIII.—Twenty-two years ; syphilis appearing at the
début of pregnancy ; multiple secondary accidents ; treatment
of two to three months ; accouchement at term ; child still-
born.

CASE LXIX.—Nineteen years ; appearance of secondary accidents in the third month of pregnancy ; no treatment ; accouchement at seven months ; child still-born.

CASE LXX.—Twenty-six years ; indurated vulvar chancre appearing at third month of pregnancy ; treatment of from two to
three months ; accouchement at seven and a half months ; child
still-born.

CASE LXXI.—Twenty years ; pregnancy of five months ; début of secondary syphilis unknown ; vulvar syphilides ; syphilitic
fever ; costal periostitis ; treatment of a few months ; accouchement at term ; child still-born.

CASE LXXII.—Twenty-one years ; contagion at third month
of pregnancy ; indurated chancre of the vulva ; vulvar syphilides ;
treatment of a few weeks ; abortion.

CASE LXXIII.—Nineteen years ; début of syphilis unknown ;
papulo-squamous syphilide ; alopecia ; tonsillar syphilides ; treat-

ment for several months ; accouchement at term ; child syphilitic, dying in three weeks.

CASE LXXIV.—Twenty-six years ; début of syphilis unknown ; papular syphilide ; vulvar syphilides of circinate form ; cephalalgia ; pregnancy ; treatment not prolonged ; abortion.

CASE LXXV.—Twenty years ; pregnancy of four or five months ; début of syphilis unknown ; secondary accidents ; treatment of a few weeks ; premature accouchement ; child dying at five days.

CASE LXXVI.—Twenty-two years ; contagion in the first months of pregnancy ; vulvar, anal, and tonsillar syphilides ; cephalalgia ; mercurial treatment prolonged several months ; accouchement at term ; child healthy (lost sight of when six weeks old).

CASE LXXVII.—Twenty years ; pregnancy ; début of syphilis unknown ; pustulo-crustaceous syphilide ; no treatment ; accouchement at seven months ; child dying when five days old.

CASE LXXVIII.—Twenty-one years ; syphilis dating from eighteen months ; mercurial and iodide treatment quite regularly followed and long protracted ; accouchement at term in the second year of the disease ; child living and healthy.

CASE LXXIX.—Twenty-two years ; début of syphilis unknown ; no treatment ; abortion at two months.

CASE LXXX. — Twenty-eight years ; syphilis dating from eleven years ; treatment very insufficient ; three pregnancies since the début of the disease ; three abortions—at six weeks, at six months, at seven months.

CASE LXXXI.—Twenty-four years ; début of secondary syphilis unknown ; vulvar syphilides ; cephalalgia ; no treatment ; abortion at two months.

CASE LXXXII. — Twenty-two years ; secondary accidents making invasion in the course of pregnancy ; cutaneous and mucous syphilides ; mercurial treatment of several months ; accouchement at term ; child syphilitic, dying at two months.

CASE LXXXIII.—Twenty-five years ; pregnancy of three to

16

four months ; début of syphilis unknown ; roseola ; palmar psoriasis ; vulvar and peri-vulvar syphilides ; cephalalgia ; neuralgic pains ; febrile attacks ; treatment of a few months ; premature accouchement ; child still-born.

Case LXXXIV.—Nineteen years ; pregnancy ; début of syphilis unknown ; confluent vulvar syphilides ; papulo-squamous syphilide ; cephalalgia ; nervous troubles ; no treatment ; accouchement at seven months ; child still-born.

Case LXXXV.—Thirty-five years ; début of syphilis not known, certainly dating back from a distant period ; multiple gummous tumors ; five pregnancies ; four children dying, all soon after birth ; last child surviving.

Case LXXXVI. — Twenty-two years ; pregnancy of seven months ; secondary syphilis, début unknown ; cutaneous syphilides ; vulvar and buccal syphilides ; analgesia ; treatment, nature unknown, regularly followed during several months ; accouchement at term ; child dying on fifteenth day.

Case LXXXVII.—Twenty-two years ; pregnancy of four to five months ; syphilis dating from fourteen months ; vulvar syphilides ; treatment of some months ; accouchement at term ; child dying at twenty days.

Case LXXXVIII.—Eighteen years ; accidents of secondary syphilis coinciding with début of pregnancy ; treatment of one month ; papulo-hypertrophic syphilides of the vulva ; accouchement at term ; child still-born.

Case LXXXIX.—Twenty-four years ; secondary syphilis, début unknown ; no treatment ; abortion.

Case XC.—Twenty-seven years ; secondary accidents appearing toward the sixth month of pregnancy ; no treatment ; accouchement at term ; child still-born.

Case XCI.—Nineteen years ; secondary syphilis, début unknown ; no treatment ; abortion at five months.

Case XCII.—Twenty-eight years ; secondary syphilis, début unknown ; pregnancy of five months ; no treatment ; abortion.

Case XCIII.—Twenty-three years ; pregnancy of six months ; secondary syphilis, début unknown ; vulvar and anal syphilides ; treatment of some months ; accouchement at term ; child stillborn.

Case XCIV.—Thirty-three years ; pregnancy of three months ; vulvar syphilides ; papulo-squamous syphilide ; palmar psoriasis ; no treatment ; abortion at six months.

Case XCV.—Twenty-two years ; secondary accidents appearing toward the fifth month of pregnancy ; vulvar, peri-vulvar, and anal syphilides, etc.; palmar psoriasis ; alopecia ; treatment of a few months ; accouchement at term ; child living, lost sight of when fifteen days old.

Case XCVI.—Nineteen years ; secondary syphilis, début unknown ; ulcerating syphilides of the vulva ; palmar psoriasis ; no treatment ; abortion at six months.

Case XCVII.—Twenty years ; syphilitic chancre of the vulva in the fifth month of pregnancy ; roseola ; buccal syphilides ; interdigital syphilides ; cephalalgia ; mercurial treatment of four months ; accouchement at term ; lost sight of after four or five weeks.

Case XCVIII. — Twenty-four years ; pregnancy of three to four months ; début of syphilis unknown ; vulvar, anal, and buccal syphilides ; treatment of some months ; accouchement at term ; child syphilitic ; lost sight of when two months old.

Case XCIX. — Twenty-two years ; pregnancy of about five months ; syphilis dating from one year ; vulvar, peri-vulvar, and anal syphilides ; treatment of several months ; accouchement at term ; child syphilitic, surviving.

Case C.—Twenty-eight years ; contagion during pregnancy ; vulvar syphilides ; alopecia ; treatment not prolonged ; accouchement at term ; child dying at two months.

Case CI.—Twenty-three years ; secondary syphilis, début unknown ; no treatment ; abortion at two and a half months.

Case CII.—Twenty-nine years ; secondary syphilis appearing

in the second half of pregnancy ; papulo-squamous syphilides ; buccal and vulvar syphilides ; febrile attacks ; mercurial treatment until the end of pregnancy ; accouchement at term ; child syphilitic; treated, surviving.

Case CIII. — Twenty years; secondary syphilis, début unknown ; vulvo-anal and buccal syphilides; treatment, nature unknown ; accouchement at term ; child surviving.

Case CIV.—Twenty-one years ; accidents of secondary syphilis appearing in the second half of pregnancy ; vulvar and buccal syphilides ; treatment of fifteen days ; accouchement at term ; child still-born.

Case CV.—Twenty-four years ; pregnancy of eight months and a half ; début of syphilis unknown ; palmar psoriasis ; treatment for several months ; child healthy, surviving.

Case CVI.—Twenty years ; pregnancy of three months ; indurated chancre of the vulva ; papular syphilide ; genital syphilides ; febrile attacks ; treatment of from two to three months ; abortion.

Case CVII.—Twenty-one years ; syphilis dating from eight months ; pregnancy of four months; erythemato-papular syphilide ; vulvo-anal syphilides ; treatment of ten days ; abortion at five and a half months.

Case CVIII.—Twenty-three years; pregnancy of three months; syphilis dating from two years ; papular syphilide ; vulvar and buccal syphilides ; irregular treatment ; accouchement at eight months ; child dying in seventeen hours.

Case CIX.—Twenty-two years ; syphilitic chancre of the neck of the uterus in the seventh month of pregnancy ; mercurial treatment ; accouchement at term ; child healthy (at least up to its exit from the hospital, six weeks old).

Case CX.—Twenty years ; infected from the début of marriage ; indurated chancre of the lip ; roseola ; vulvar and buccal syphilides; cephalalgia ; neuralgic pains ; treatment of several months ; accouchement at term (ten months after marriage); child syphilitic, dying at one month.

CASE CXI.—Twenty-three years ; début of pregnancy and secondary accidents of syphilis ; roseola ; buccal syphilides ; treatment very irregular ; accouchement at eight months ; child still-born.

CASE CXII. — Twenty-five years ; secondary syphilis, début unknown ; vulvar syphilides ; palmar psoriasis ; pregnancy of four months ; treatment of six weeks ; abortion.

CASE CXIII. — Twenty years ; contagion toward the third month of pregnancy ; syphilitic chancres ; roseola ; cephalalgia ; nervous phenomena ; analgesia ; peripheric algiditis ; losses of consciousness ; febrile attacks ; mercurial and iodide treatment, very active and prolonged during the entire pregnancy ; accouchement at term ; child healthy ; six months after accouchement the mother presented some papulo-circinate syphilides upon the legs.

CASE CXIV.—Twenty years ; pregnancy ; début of syphilis unknown ; cutaneous and mucous syphilides ; treatment of some weeks ; abortion at three months ; one year later second pregnancy ; child syphilitic, dying at five months.

CASE CXV.—Twenty-two years ; pregnancy of six to seven months ; début of syphilis unknown ; vulvar and buccal syphilides ; roseola ; analgesia ; two or three months' treatment ; accouchement at term ; child syphilitic ; treated, surviving ; contamination of the nurse by the child ; syphilis of the nurse very severe.

CASE CXVI.—Twenty-two years ; chancre of the breast transmitted by a syphilitic nursling ; cutaneous syphilides ; vulvar and buccal syphilides ; treatment of some weeks ; pregnancy a few months later ; accouchement at term ; child syphilitic, dying at five weeks.

CASE CXVII.—Nineteen years ; infected in the first months of marriage in the course of pregnancy ; treatment of a few weeks only ; abortion.

CASE CXVIII.—Twenty-one years ; syphilis by conception ; secondary accidents ; treatment of some weeks ; accouchement at term ; child probably syphilitic ; dying of convulsions when one month old.

Case CXIX. — Thirty-one years ; two children living and healthy, born before the contagion ; contagion in the course of the third pregnancy ; treatment of some weeks at its début, and since then no medication ; child still-born ; afterward four pregnancies, from year to year ; three giving children still-born, or dying after a few days ; of the fourth only, child surviving, feeble, but appearing to have never been affected with specific accidents.

Case CXX.—Twenty-eight years ; contagion one year after marriage ; multiple secondary accidents ; cutaneous and mucous syphilides ; crusts of the hairy scalp ; alopecia ; neuralgic pains ; cephalalgia, etc. ; treatment insignificant ; six pregnancies in four years ; six abortions.

Case CXXI.—Twenty-nine years ; pregnancy of five to six months ; début of syphilis unknown ; secondary accidents ; treatment of some weeks ; accouchement at eight months ; child still-born.

Case CXXII.—Seventeen years ; syphilis recent, and pregnancy of two or three months ; vulvar, anal, perineal, and tonsillar syphilides ; crusts of the hairy scalp ; no treatment ; abortion a few days after entrance into hospital.

Case CXXIII.—Twenty-one years ; pregnancy of five months ; début of syphilis unknown ; papular syphilide ; vulvar and buccal syphilides ; treatment of two to three months ; accouchement almost at term ; child syphilitic, dying at three weeks.

Case CXXIV.—Thirty years ; chancre in the fourth month of pregnancy; cutaneous syphilides; alopecia; vulvar syphilides; treatment of some weeks ; accouchement before term ; child still-born.

Case CXXV.—Twenty-seven years ; début of syphilis unknown ; palmar psoriasis ; tonsillar syphilides ; vulvar and perineal syphilides ; alopecia ; treatment very irregular ; four pregnancies in two years succeeding invasion of syphilis ; four abortions from two to four months.

CASE CXXVI.—Twenty-five years ; syphilis dating from seven months ; pregnancy of five months ; cutaneous syphilides ; no treatment ; abortion two days after entrance into hospital.

CASE CXXVII.—Twenty-two years ; syphilis contracted at beginning of marriage ; no treatment ; abortion at three months ; secondary accidents ; papular syphilide ; buccal syphilides ; vulvar syphilides ; treatment of a few weeks ; the following year, abortion at five months.

NOTE IV.

THE ordinary period of syphilitic incubation some-times exemplifies the curious fact of a syphilis which, contracted *before* marriage, does not make its first inva-sion until *after* marriage.

Cases of this kind are naturally quite rare. But, nevertheless, I have the records of four which are quite authentic. The following will serve as an example:

M——, aged twenty-eight years; good constitution; had typhoid fever at the age of fourteen years. Nothing else, besides this, than passing indispositions. In the way of venereal accidents, had two blenorrhagias, at twenty-two and twenty-four years of age; perfectly cured.

Fifteen days before the date fixed for his marriage, M—— gave a large supper to his friends, under the pretext of bidding adieu to bachelor life. Stupefied by copious libations, he allowed himself to be induced to spend the night with a former mistress. This woman was at the time under treatment for accidents of secondary syphilis, and still presented certain "*boutons*" on the vulva, which her physician had characterized (I subse-quently learned) as mucous patches. Afterward, I had occasion to see this patient several times, and observed undoubted accidents of syphilis upon her.

M—— marries in a perfect condition of health. Fifteen days after his nuptials, he observes upon the glando-preputial furrow a slight redness, somewhat erosive. He gives it no further attention, believing it to be "an excoriation from intercourse with his wife." His sexual relations are not interrupted. Still, the erosion exists; it becomes enlarged, and seems to become tumefied at its borders. Cauterization with *vinaigre de Bully*, and continuation of his sexual relations. Some days later only does M—— become uneasy, and consult his physician, who expresses to him the liveliest fears as to the nature of the accident. Frightened, he rushes to me, and I note the following condition: Upon the glando-preputial furrow a superficial erosion, oval-shaped, of the diameter of a lentil; surface smooth, reddened, gray in the center, lardaceous, pseudo-membranous; borders adherent, slightly elevated; base renitent, hard, of a dry hardness, almost characteristic. A single ganglion in the corresponding groin, hard and indolent. I confirm the diagnosis of my *confrère*, and I consider myself authorized in declaring to the patient that the lesion with which he is affected is a *syphilitic chancre*, resulting from a contagion dating back several weeks.

The following days, the characteristics of this lesion become more accentuated. The sore extends, and the induration, especially, becomes exuberant, cartilaginous. Several glands become implicated, constituting a veritable inguinal pleiad. The syphilitic infection is then absolutely manifest.

Not until this time does the patient avow to me his experience, and bring the woman to me with whom he had connection several days before his marriage. The verification of the syphilis in this woman and the details which she gives in regard to her disease complete and confirm the diagnosis of my patient's lesion.

Six weeks later, the patient's body is covered with a roseola. Afterward, tonsillar syphilides, crusts of the hairy scalp, slight alopecia, cervical adenopathies; mercurial treatment; disappearance of the accidents.

The wife of this patient would not at first consent to an examination which her husband had suggested under some pretext or other. Consequently, I did not see her until about two months and a half after her marriage. At this time there remained no trace of vulvar accidents. But the patient stated that she had had a slight "excoriated pimple" some weeks previous, which had produced "some swelling" of the labia; and on examination there was found in the groin, on the same side as this lesion, a ganglionic pleiad very markedly accentuated, and leaving scarcely a doubt of a syphilitic infection of recent date.

A fortnight later, the patient complained of general lassitude, headache, vague pains in the limb. Then, a roseola soon manifested itself, which dissipated all uncertainty as to the situation.

Subsequently, palmar psoriasis, tonsillar syphilides, alopecia.

To recapitulate, then:

1. Fifteen days *before* marriage, connection with a woman affected with vulvar syphilides.

2. Marriage in apparently perfect condition of health.

3. Fifteen days *after* marriage, appearance of a syphilitic chancre, followed by secondary accidents, after the normal incubation.

4. Contamination of the young wife from the chancre of the husband, the nature of which chancre was unrecognized at its origin.

My other three cases are, so to speak, reproductions of the one just narrated. They all relate to chancres breaking out *after* marriage, as a consequence of a contagion preceding the marriage from eight to seventeen days. Three times in four, the young wives contracted the contagion, and the fourth only escaped, thanks to the accident of a quite prolonged indisposition, which suspended all intercourse. In every case, in fine, the long duration of the incubation deceived the husbands as to the nature of the accidents, and exposed them to the risk of infecting their wives.

NOTE V.

THE following case is interesting in two respects: On
the one hand, it shows in a general way what may be the
consequences of a premature marriage in syphilis; on
the other hand, it illustrates the dangers incidental to the
raising of a syphilitic infant, when this infant has been
committed to a nurse instead of being suckled by its
mother.

I. N—— contracts syphilis. At first he is treated
by a pharmacist, who gives him pills of a "secret" com-
position. Four months later he comes to consult me,
and I observe the following accidents upon him: papular
syphilide covering the thorax and the limbs; tonsillar
syphilides; crusts of the hairy scalp; alopecia; cervical
adenopathies.

Mercurial treatment. Disappearance of the accidents
in a few weeks.

Afterward recurrence of a papulo-squamous syphilide,
affecting the scrotum. Mercurial treatment is resumed;
later, iodide of potassium.

The patient is treated regularly during five or six
months, after which I lose sight of him. I have since

learned from him that, believing himself cured, he no longer followed any medication after this time.

Two years after the début of his syphilis, he marries, without taking counsel of me or of any other physician. He was, nevertheless, far from being cured at this time, as was demonstrated by the reappearance of various accidents in the following years : cutaneous syphilides, buccal erosions, onyxis, periostosis, etc.

II. Some months after his marriage, the wife of N—— commenced to complain of neuralgias of the head, of intense pains in the limbs, of insomnia, of general *malaise*, of febrile attacks, etc. These various symptoms were at first treated with the sulphate of quinine, but without success. Soon a confluent eruption covered the body, and enlightened the physician as to the nature of the anterior accidents which had, until then, resisted his medication.

At this time this woman was brought to me, and I noted upon her accidents of a nature incontestably syphilitic : generalized papulo-squamous syphilide, palmar psoriasis, crusts of the hairy scalp, with scattered alopecia, tonsillar erosions, cervical adenopathies, etc. In addition, about this same date, she became pregnant.

Mercurial treatment : rapid disappearance of the accidents ; accouchement, at term, of a fine child, which, contrary to my express recommendations, was committed to a nurse at some distance from the city.

III. I had lost sight of these two patients for a certain time, when, one day, N—— summoned me to his house, to present to me at the same time—1. His diseased infant ; 2. The nurse of this infant infected by it ; 3. The husband of this nurse infected by his wife. And, in fact, a long interrogation, followed by a careful examination revealed to me the following series of results :

1. The infant remained exempt from every morbid symptom during the first four or five weeks. After that time its body was covered with pimples, especially around the buttocks, its mouth was ulcerated, its nose "ran profusely." It became thin, emaciated, and they had fears for its life during several months. Nevertheless, it grew better, thanks to the treatment which was prescribed for it by the physician of the locality (mercurial frictions, baths of the sublimate, iodide of potassium, etc.). It now presents various specific accidents : erosive syphilides at the buccal commissures ; papulo-ulcerous syphilides of the margin of the anus, etc.

2. Some weeks after the invasion of these accidents upon the nursling, the nurse's breast became "ulcerated." She was not told the nature of the lesion on her bosom, but she says that she was treated with mercurial pills at this time. Besides, some weeks later, she suffered from sore-throat and an inflammation of the vulva, with "excoriated pimples" ; her body was covered with a red eruption, and her hair fell out to such an extent that she "thought she would become absolutely bald." I observed upon her, at the time of my visit, undoubted traces of a squamous syphilide, cervical adenopathies, a severe alopecia, and pigmentary macules scattered upon the neck.

3. The infant of this woman (which she nursed at the same time as the infant of N——) had been healthy from its birth, and it had continued to "thrive" during several weeks. But two months ago it suddenly commenced to waste away. Its body became covered with a papular eruption, its mouth became ulcerated, its legs became swollen ; it then died in a state of consumption. The at-

tending physician, I was informed, had no doubt that the child succumbed to a syphilis contracted after its birth.

4. Finally, the husband of this woman, a man of regular habits and of irreproachable morality, became diseased some months after his wife. He commenced by presenting several "*boutons*" upon the penis, then he was affected with a confluent eruption, with pains in the head, with sore-throat, etc. I found him, at the time he came under my observation, in an active condition of secondary syphilis: erythemato-papular syphilide, eruption of crusts of the hairy scalp, cervical adenopathies, buccal syphilides, etc. In addition, I found, on the glando-preputial furrow, two cicatricial indurations with a double inguinal pleiad, which were unquestionably the remains of the primitive infection.

To recapitulate, then:

1. Premature marriage of a syphilitic subject.

2. Contagion transmitted from the husband to his wife.

3. Birth of a syphilitic child, which is committed to a nurse, in defiance of medical advice.

4. Contagion transmitted by this child to its nurse.

5. Contagion transmitted by this nurse to her own child, which becomes emaciated, wastes away, and dies.

6. Contagion transmitted by this same nurse to her husband.

That is to say, five cases of syphilis and one death resulting from the premature marriage of a man with a syphilis not yet extinct.

NOTE VI.

FATHER SYPHILITIC; CHILD SYPHILITIC; MOTHER SEEM-
ING AT FIRST EXEMPT, BUT PRESENTING, SIX YEARS
LATER, AN ACCIDENT OF TERTIARY SYPHILIS.

"In 186- I had under my care Madame Z—— during
the last six months of her pregnancy, which, moreover,
was complicated with frequent nausea and occasional
vomiting. The 6th of April this woman gave birth to a
little girl, well developed, healthy in appearance, weighing
three kilogrammes one hundred and fifty-two grammes.
The delivery was completed naturally, twenty minutes
after the accouchement. The placenta was healthy. The
after-results of the accouchement were favorable.

"About the tenth day the child had a little fever,
green stools, erythema of the buttocks. On the fifteenth
day an eruption showed itself on different portions of the
surface. It soon took on the characteristics of a syphilitic ecthyma. The 15th of May, mucous patches, as
manifest and as typical as possible, appeared around the
anus and on the vulva. A treatment, consisting of baths
of the sublimate and frictions with mercurial ointment,
soon caused the disappearance of these various accidents.

"Nevertheless, the mother continued, and has since
continued, to nurse her child. She has not ceased to be
perfectly well, and, notably, she has never presented any
symptom which could be attributed to syphilis. I will

add that a careful interrogatory, as minute as possible, into the antecedents of this woman did not enable me to discover anything specific in her history. When she ceased nursing her infant (fourteen months later), she was a little anæmic, and complained of a persistent pain between the shoulders. She improved rapidly, and without any medication, as soon as she ceased nursing.

" The father, interrogated by me as to his antecedents, had confessed to me that he had contracted an infecting chancre four months before his marriage, and that even at the moment of conception he was still affected with various secondary accidents (mucous patches at the anus, mucous patches on the tonsils, and disseminated crusts of the hairy scalp).

" I found myself in the presence then—1. Of a syphilitic father, still presenting syphilitic accidents at the moment of the conception of his child. 2. Of a syphilitic child commencing to present undoubted manifestations of syphilis on the fifteenth day after its birth. 3. Of a mother not infected, and appearing to have never suffered any specific accident before her accouchement, and having nursed her syphilitic infant during fourteen months without contracting the least contagious symptom from it.

" This fact overturned all my beliefs upon syphilitic heredity, and notably my cardinal conviction, viz., that, if a child be born tainted with syphilis, the mother must certainly have been infected. In my opinion, as I had established in a previous memoir, *pas de syphilis de l'enfant sans syphilis de la mère.*

" During six years, I was privileged to look after the health of this family, and I can testify that—1. The child, submitted to the above-mentioned treatment, has survived.

17

Although somewhat lymphatic, it has never ceased to enjoy very good health. 2. The father, who has followed a rigorous and prolonged treatment, has presented no more specific manifestations. 3. The mother, scrupulously observed, has not ceased to continue well, apart from certain passing indispositions.

"I confess that this case (followed up by me from day to day, so to speak) very much disturbed my former convictions. I was even preparing to publish it, when, in the month of October, 186-, Madame Z—— came to consult me in regard to a tumor on her right arm. This tumor, situated beneath the skin, immediately above the olecranon, was of the volume of a pigeon's egg. Its circumference and base were hard, but its central portion was soft. It had never given rise to pain, and it still remained indolent, even on palpation and pressure. The integument around the portion in process of softening presented a brownish-red color. I examined this tumor with great care, and I arrived at the conclusion that it could only be constituted by a *syphilitic gumma.* Ten days later, an abscess formed in the tumor, and opened at its central portion. It gave exit to a liquid which was composed of two distinct portions, one transparent and gelatinous, similar to dissolved gum, the other purulent. The tumor once opened, I perceived that the base of the abscess was grayish, as if putrilaginous. The opening rapidly enlarged, presenting a sinuous contour, with borders neatly and perpendicularly cut. The base of the tumor still remained indurated. These characteristics, this evolution, tended to confirm me in my first impression. I diagnosticated a gummous tumor, and I do not indeed believe that any other diagnosis could have been arrived at.

"Treatment by iodide of potassium, doses at first in-

creasing, then decreasing ; rapid improvement of the lesion ; cure in less than three weeks.

"The evidence was, then, conclusive. Besides, what
other diseases could this tumor be confounded with ? With
an anthrax ? With an abscess ? But the absence of pain,
the absence of inflammatory phenomena, the objective
aspect of the lesion, the morbid evolution, all exclude
such hypotheses ; and, more, the rapid cure effected by
the iodide of potassium completely demonstrates, at least
in my opinion, that it was a lesion of tertiary syphilis.

"It is certain, then, that Madame Z—— had been infected at some previous date, either before or during her
pregnancy. Her syphilis had been obscure, fugacious.
This syphilis had passed unrecognized both by the patient
and myself. Finally, it did not reveal itself in a manifest
shape until *six years later*, by the outbreak, quite unexpected, of a lesion of the tertiary order.

"Altogether, this case, which at first seemed as if it
must of necessity overturn the theory which I for a long
time had sustained (in accord with that of M. Cullerier,
M. Notta, and other observers), viz., that *every syphilitic
child is born from a syphilitic mother*—this case, I say,
has, on the contrary, furnished an additional argument
for the theory in question, and confirms it absolutely."—
(Dr. A. Charrier.)

NOTE VII.

A CARDINAL fact which I have endeavored to place in strong relief in this work relates to the most serious of the hereditary consequences of paternal syphilis.

I have asserted, and I think demonstrated, that the child procreated by a syphilitic father is very often stamped with a sort of *inaptitude for life.* In other words, this child is liable to die early, either *in utero,* or within a short time after accouchement.

I can not reproduce here all the facts which have served to establish my conviction upon this point; but I think that I ought to place a certain number under the eyes of the reader as illustrative cases :

CASE I.—Indurated chancre of the penis ; roseola ; palmar psoriasis ; tonsillar syphilides ; treatment from six to eight months, but irregularly followed ; marriage five years after the début of the syphilis ; wife perfectly healthy, remaining absolutely exempt ; four pregnancies ; four abortions ;* at this time

---

* I specify, once for all, that in this case, as in the following cases, abortion or premature accouchement could not possibly have been due to any cause, either accidental or constitutional, depending upon the wife. We have here to deal only with cases (I have selected them designedly) where after careful examination, after searching for and excluding every other cause, the death of the fœtus remains exclusively imputable to the syphilis of the husband.

the patient is subjected to a new treatment (mercury and iodide of potassium for about a year) ; four pregnancies afterward ; four accouchements at term ; children living and healthy (the eldest is at present twelve years old).

CASE II.—Indurated chancre ; buccal syphilides ; insignificant treatment of a few weeks' duration ; married fifteen months after début of syphilis ; wife remaining absolutely exempt ; nine pregnancies ; five *abortions ;* three *accouchements before term*, infants living from a few hours to three days ; ninth pregnancy at term, resulting in a living infant, which, fifteen days later, was covered with syphilides.

CASE III.—Indurated chancre, followed by some secondary accidents ; treatment from six to eight months ; ocular paralysis three years later ; marriage seven years after the début of the syphilis ; tertiary accident (exostosis) the same year ; treatment resumed with vigor, wife remaining exempt ; four pregnancies from year to year ; first pregnancy terminated by abortion ; second pregnancy : accouchement at eight months of a dead infant ; third pregnancy : accouchement at term, infant dying in a few hours ; fourth pregnancy : accouchement at term, child surviving and healthy.

CASE IV.—Indurated chancre ; cutaneous and mucous syphilides ; treatment for about a year ; marriage four years later ; the first pregnancy is terminated by an accouchement at term ; child living and healthy ; two years later, reawakening of the diathesis ; tuberculo-ulcerative syphilide, rebellious and recurring ; sclerous glossitis ; gumma ; ecthyma ; this crop of accidents is prolonged during three years, despite an energetic treatment ; it coincides with three pregnancies which all terminate in abortion ; wife uncontaminated.

CASE V.—Indurated chancre, followed by a few very slight secondary accidents ; mercurial treatment during two months ; the following years, small doses of iodide of potassium from time to time ; marriage fourteen years after début of syphilis, wife remaining uninfected ; two pregnancies in the course of the two years succeeding marriage ; the first terminating by accouchement at term of a dead infant ; the second producing a syphilitic infant, which dies in three weeks ; consecutively, husband has recurrence of tertiary accidents.

Case VI.—Syphilis ; indurated chancre of the penis ; two to three months mercurial treatment in feeble doses ; no secondary accidents remarked ; marriage four years later, wife remaining exempt ; three pregnancies terminated by three abortions ; at this time accidents of tertiary form ; treatment energetic, and prolonged with mercury and the iodide of potassium ; a fourth pregnancy happening a year later results in a child at term, living and healthy.

Case VII.—Indurated chancre of the prepuce ; confluent roseola ; mercurial treatment for several months ; afterward lingual syphilides ; marriage five years after the début of the disease ; wife uninfected ; three pregnancies terminating in the following manner : one abortion ; two accouchements before term, infants dead ; consecutively, the husband affected with a psoriasiform syphilide.

Case VIII.—Chancre of the lip ; cutaneous syphilides ; buccal patches ; onyxis, mercurial treatment for several months ; marriage one year later, wife remaining exempt ; three pregnancies in three years ; the first two infants still-born ; the third is born syphilitic, and dies in three months ; consecutively the husband affected with a psoriasiform syphilide of the hands.

Case IX.—Indurated chancre ; secondary accidents ; mercurial treatment of one month only ; marriage two years later, wife remaining exempt ; two pregnancies terminate by abortion ; third pregnancy, infant at term, cachectic, dying in a few weeks ; afterward husband is reattacked with ulcerative syphilides of the penis.

Case X.—Labial chancre ; roseola ; tonsillar syphilides ; mercurial treatment of three months ; marriage ten years later, wife remaining exempt ; four pregnancies ; two abortions ; two infants born at term, hydrocephalic, soon dying ; consecutively the patient is affected with cranial osteitis ; symptomatic encephalitis ; death.

CASE XI.—Indurated chancre of the prepuce ; papular syphilide ; buccal syphilides ; onyxis ; cervical adenopathies ; tibial periostosis ; treatment of six months ; marriage three years later, wife remaining exempt ; four pregnancies ; the first two terminate by abortion ; the third brings a syphilitic infant to term, which dies the second day ; resumption of treatment ; fourth pregnancy : child syphilitic, surviving.

CASE XII.—Indurated chancre of the sheath ; roseola ; anal and buccal syphilides ; treatment of four weeks by mercury, then treatment of two months with iodide of potassium ; marriage in the second year of the disease, wife remaining exempt ; two miscarriages ; third pregnancy, giving a syphilitic infant (syphilides, pemphigus, osseous lesions), which soon dies.

XIII.—Indurated chancre ; secondary syphilides of the skin and mucous membranes ; iritis ; treatment for some months ; marriage two years later, wife remaining uninfected ; four pregnancies, three quite near each other ; the first two terminate by abortion ; the third gives a syphilitic infant, which survives, thanks to an energetic treatment ; the fourth produces an infant which has never to this day presented specific accidents.

XIV.—Indurated chancre ; some secondary accidents ; treatment for several months ; marriage four years later ; wife remaining exempt ; two abortions ; third infant syphilitic, dying when four weeks old.

THE END.

www.ingramcontent.com/pod-product-compliance
Lightning Source LLC
Chambersburg PA
CBHW021520210326
41599CB00012B/1327